Home Performance Diagnostics

THE GUIDE TO ADVANCED TESTING

second edition

Corbett Lunsford and the
Building Performance Workshop

ISBN 978-0-615-59475-0

Published by the Building Performance Workshop
a division of Green Dream Group LLC
5756 N. Ridge Ave. #11
Chicago IL 60660
773.271.5310
Support@BuildingPerformanceWorkshop.com

For more information, videos, and training, visit:
www.BuildingPerformanceWorkshop.com

Design by Corbett Lunsford and John Bergman
Photos and illustrations property of the Building Performance Workshop
This edition's text is set in News Gothic MT

All Building Performance Institute (BPI) standard excerpts printed with permission of BPI, Inc.

All tools shown are either well-used property of Chicago Home Performance, a division of Green Dream Group LLC, or were generously loaned by TruTech Tools (*www.TruTechTools.com*). Any branded non-testing equipment (including HVAC equipment) is used for these real-life demonstrations, and their inclusion in the photographs does not imply an endorsement.

TruTechTools

How To Use The Book

There are a number of cool features to this book, which I hope will make it very useful to you in the field. Also, every diagnostic test you'll find in each of the four main categories (Envelope, HVAC, and Electrical, and Prerequisites) follows a precise format, for clarity and ease of implementation:

Test Number and Title →

3.06 Electric Love Test

List of tools required →

Materials Needed:

- electric love sniffer
- fans! hoses! bring everything!

Video demo links →

Video demonstration available at www.BuildingPerformanceWorkshop.com/videos

Conceptual overviews →

CONCEPT

Electric love is often thought of as un-testable, but as you can see in **Figure 3.06.1**...

Figures and Illustrations →

Figure 3.06.1

one	two
bees	pizza

Prep steps →

PREPARATION

Ensure your test subject consents to be tested. Turn on the love sniffer and let it warm up while you start the vacuum cleaner.

Best practice tips →

***TIP:** *Doing housework for your subject increases love's electricity.*

Multiple methods →

3.06-A Logical Method

3.06-B Ridiculous Method

High-definition photos →

Step-by-step instructions →

TESTING

Bring the home to –50 Pa and do all the dishes and laundry. If the electricity isn't measuring off the charts on your manoanemometer, try buying flowers.

Learning what it means →

INTERPRETATION

Use **Equation 3.06.1** to interpret the test results:

Sophisticated equations →

Equation 3.06.1 $TEST = T + EST!$

Every diagnostic also includes a QUIZ, so you can put your new skills to work!

ACKNOWLEDGEMENTS

SPECIAL THANKS TO THE FOLLOWING PEOPLE, WITHOUT WHOM THIS BOOK WOULD NOT BE IN YOUR HANDS NOW:

- My colleagues at Green Dream Group: John Bergman (editor, photographer/engineer, and my right hand man); the voice of our company, Carissa Ilg (the best eye for detail I know); and Grace Lunsford, my boss/best friend/beautiful wife.
- Dustie Lynch, my tireless proofreader, my role model as a trainer, and my awesome mom; and Paul Lynch, my favorite attorney, technical inspiration, photographer, and dad.
- Jim Bergmann and Bill Spohn at TruTech Tools, close technical advisors and generous suppliers of many of the tools you see photographed herein.
- Everyone else who gave their time and attention to this book's development, including, in no particular order:
    - Matt Tahja, Green Dream Group
    - Bronson Shavitz, Shavitz Heating and Air Conditioning
    - Vikki Murphy, BPI
    - Dennis Stroer, Calcs-Plus
    - Gary Nelson, The Energy Conservatory
    - Colin Genge, Retrotec Inc.
    - Richard White, Mobile HVAC Training Systems
    - Larry Armanda, Therma-View Infrared and Energy Consultants
    - Scott Suddreth, Building Performance Engineering
    - Brett Dillon, IBS Advisors
    - John Straube, Building Science Corporation
- All the people I've learned this trade from, including Jim Cavallo, Brian Kumer, Mark Jansen, Keith Williams, and all those who have agreed to answer my never-ending questions on the Building Performance Podcast.

TABLE OF CONTENTS

FOREWORD

Our buildings, including over 110 million households in the United States, are among the nation's largest investments, and important to the safety, health, productivity and happiness of the people who live and work in them. Experience has shown that our housing is often built less than perfectly, and needs on-going maintenance and updating. The rising cost of energy, heightened expectations of comfort, and concern for environmental impact is driving unprecedented changes in the demand for housing performance.

In the last fifty years, a small cadre of North American researchers has applied physical science to the understanding and prediction of building performance, an area of knowledge now known as building science. Meanwhile, a somewhat larger group of engineers, architects, builders, and designers began to provide practical feedback and observations. Today and in the future, we will need a small army of hands-on professionals in every state, town, and suburb to visit, assess, measure, and understand the performance of our existing buildings. *Home Performance Diagnostics* is a timely repository of very practical, hands-on, and useful information about how that army can use testing and diagnosis to understand and improve the performance of homes.

Lord Kelvin, one of the most influential scientists of the last millennium, said:

"When you can measure what you are speaking about, and express it in numbers, you know something about it..."

The methods of testing and measuring outlined in this book would convince Lord Kelvin that we, as an industry, know something about building performance.

Dr. John Straube, P.Eng.
Associate Professor, Department of Civil Engineering and School of Architecture, University of Waterloo, Waterloo, Canada
Principal, Building Science Corporation, Somerville, Massachusetts

INTRODUCTION

0.01 What's In The Book

I wrote this book for the same reason you're reading it: people in our profession need a handy reference guide for home performance testing. There are books on building science, there are guides to building and retrofitting homes, and there are equipment manuals; but there hasn't been a step-by-step reference guide for hands-on home performance testing until now.

At our training center in Chicago, I've seen contractors, architects, HVAC technicians, plumbers, engineers, building inspectors, and newcomers alike struggle to master a dozen or more complicated pieces of equipment, each with a variety of uses, before being unleashed on the home performance industry. We give them an expectant smile after the weeklong workshop, and they go to work.

Hopefully I get a call from them soon after that, frustrated and challenged by something that 'wasn't in the books'. This is one of my many attempts to correct that, and I hope it works for you!

With this book, you can quickly learn powerful procedures for scientifically pinpointing opportunities for improvement in a home, or for verifying a job well done. I've included practical, detailed step-by-step photos for each diagnostic, and best practice tips for avoiding pitfalls.

This book is not meant to be about building science theory, making recommendations for improvement, or construction techniques. There are already quite a few well-written books on these subjects, and hopefully you own them as well. This is not one of the dozens of equipment manuals you probably own, which often are either much too short, much too long, or bizarrely translated.

You'll find overarching categories for all the testing you might want to perform (Envelope, HVAC, and Electrical), with a few important basic skills and concepts in the last Prerequisites section. Some diagnostics fit into multiple categories, and I hope the way that we've organized things will make it easy for you to find what you're looking for. A diagnostic might have several different possible methods (for example, 2.01 Total System Airflow has five), and each diagnostic finishes with a quiz that you can complete right here in the book, including questions for discussion and critical thinking with your fellow professionals.

As someone who has experienced starting a business focused in building diagnostics, I vividly remember the many times I found myself in a home, in front of a client or a contractor, and praying madly that all my test equipment would show me what I wanted to see. Sometimes I'd be performing a diagnostic for the first time while my client was watching me! And when surprises happened, I put on my poker face, packed up, shook hands with my client, and rushed to call my mentors to see what I had done wrong. Then I did what you have probably done: I called my client back, said my equipment had been 'calibrated strangely at the service center' and made a return visit to test again, at no charge.

Even though making lemonade can be delicious, I would love to spare you the stress of getting lemons like this dumped in your lap. You deserve to have the information you need at your fingertips, and to be shown exactly how certain test procedures should work. I wish you a long, successful and rewarding career in making homes more awesome!

Looking Forward,

Corbett

LIABILITY DISCLAIMER

OK, just so we're all clear that this is serious business:

This book is NOT a replacement for formal training, and many of the tests demonstrated herein can be potentially hazardous to life and property. There are many things that may go wrong with testing in the infinite variety of homes which exist, and to even attempt describing them all in this book would be totally insane. If you have not pursued hands-on training with a reputable instructor, I urge you to do so before accidentally imperiling yourself or your clients.

Especially in the case of testing combustion appliances (*see 2.05 through 2.09 and 2.15*), electrical systems (*see 3.01 through 3.05*), and envelopes (*see 1.01 through 1.05*), your safety and the safety of the home's occupants is of critical importance. The interpretations of test results and recommended next steps are only summarily covered in this book, and you should remember to always work within the limits of your abilities. If test results point to a potential safety hazard, you should take action. If you're not exactly sure what these safety hazards might look like, you should seek training. By no means attempt any test that has a **WARNING** on it if you're not experienced.

Though I've done my best to clearly demonstrate every diagnostic in this book, the Building Performance Workshop makes no guarantees about the implementation or interpretation of these tests in the field. Know your equipment, know your limits, and be safe!

0.02 The Four Aspects of Home Performance

At the Building Performance Workshop, we always strive to distill complex concepts and practices to their most essential and basic elements. This does two things:

- It enables the professional to keep a cool, clear head when faced with complicated challenges in the field
- It helps frame the way you talk to your clients about your work (clients, professional or otherwise, tend to tune out when we get technical)

In the interest of keeping it simple, therefore, we've concentrated on defining Home Performance in four aspects:

- **Heat Flow** (which everyone thinks they understand, but few truly do)
- **Air Flow and Pressure** (two sides of the same coin, which few people even attempt to grasp)
- **Moisture** (it kills the house more effectively than almost anything)
- **Air Quality and Safety** (your work's most important results are healthy, happy clients)

This list is in *increasing* order of importance. You'll notice that the two aspects which have to do with energy efficiency are at the top, which is the less important end of the list. This is why your title is better defined as a Home Performance Expert than an Energy Efficiency Expert. People will continue to call it 'Energy Auditing' when you're testing, but if you can keep reminding them that you're doing so much more than saving energy, you make yourself more valuable.

If you ever find yourself searching for a simple definition for what a 'High Performance Home' is, look no further than this:

In a high performance home, heat flow, air flow and pressure, moisture, and air quality are all perfectly controlled.

Control is the goal, just like with your car or your body. We all know how uncomfortable uncontrolled events in our cars or our bodies can be; a home is no different.

0.03 Inspection First

Before you whip out your tools and get to it, always remember that your most valuable tool is your own experience and your senses. There are 3 stages to any really valuable activity in this field:

- **Inspection**
- **Diagnostics**
- **Recommendations for Improvement**

What you see, hear, smell, and feel in the home (and taste, I suppose, but please don't let your client see you) is often the most valuable testing you perform, and it can tell you a lot about what you expect to find in the diagnostic stage.

To apply the 4 aspects in your daily work of inspection, simply look at any building component through the lens of each of the aspects. For example, a gas-fired water heater:

- *How can I better control heat flow in this component?*
    - Maximize efficiency of the heat exchanger
    - Conserve energy by carefully setting the tank thermostat
    - Ensure insulation of the storage tank
    - Ensure insulation of the distribution system

- *How can I better control air flow and pressure in this component?*
    - Ensure adequate fresh air for the combustion process
    - Test for proper venting of the combustion gases

- *How can I better control moisture in this component?*
    - Inspect for water leakage

- *How can I better control air quality around this component?*
    - Again, test for proper venting of the combustion gases

I'll include a brief rundown of the inspections that should precede the diagnostics in this book, but there's no replacement for experience, so continue to keep your senses open whenever you're in a building, on or off the clock.

1 ENVELOPE DIAGNOSTICS

1.01 Blower Door Testing

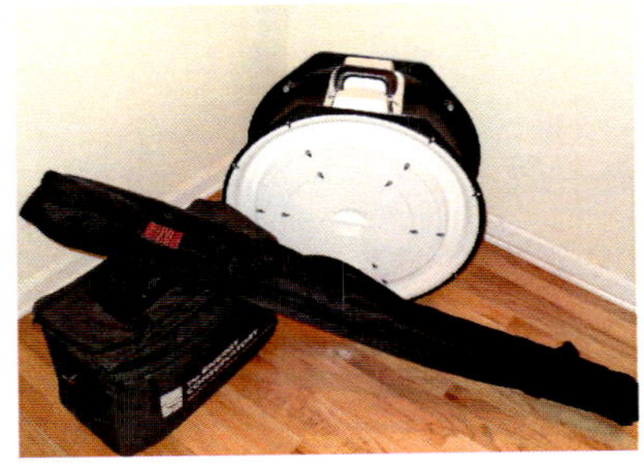

Materials Needed:

- blower door kit (complete with fan + controller, manometer, pressure hoses, frame and shroud)

CONCEPT

The blower door test is the standard test of air leakage through the home's envelope. By closing all the intentional openings in the envelope (windows, doors, mechanical dampers), and putting the home under positive or negative pressure of 50 Pascals (Pa) with respect to outdoors, you can reveal the amount and location of all the unintentional leakage through the envelope with reasonable accuracy. **WARNING:** This test can endanger personal safety and property. Do not attempt this without the proper training.

It's important to know what airflow (in cubic feet per minute at 50 Pa, or CFM_{50}) you expect to see through the fan (and, by extension, through the envelope). In order to calculate the expected airflow you must first measure the defined envelope and calculate the volume of conditioned space (*see A.01 for 'Defining the Envelope', and A.06 for 'Dimensions and Geometry'*).

If you're testing a newly constructed home, energy code generally requires a total air changes per hour during the blower door test (ACH_{50}) of less than 7, 5, or 3. That means that while the blower door is running at 50 Pa, the airflow through the fan must be low enough to replace all the air in the home less than 7 times, 5 times, or 3 times per hour.

To calculate the maximum CFM_{50} according to these limits, multiply the home's volume by the corresponding Multiplier from **Figure 1.01.1**:

Figure 1.01.1

ACH_{50} Requirement →	7	5	3
Maximum CFM_{50} Multiplier	.1167	.0833	.0500

Example:

Home Volume: 2500 ft^2 home plus 800 ft^2 basement = 3300 ft^2 total floor area

3300 ft^2 x 8 ft ceilings throughout = 26,400 ft^3

Code Air Leakage Limit: 5 Air Changes per Hour at 50 Pa = .0833 Multiplier

Maximum CFM_{50} = 26400 x .0833 = 2199 CFM_{50}

If you're testing an existing home for a retrofit or home improvement project, a different standard generally applies, as established by BPI national standards (called the Building Airflow Standard, or BAS). For existing homes, first locate the home to be tested on the map in **Figure 1.01.2**, and determine its climate zone:

Figure 1.01.2

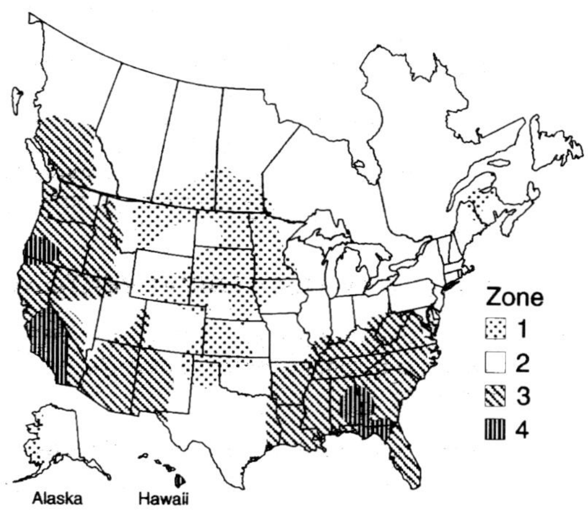

Next, using **Figure 1.01.3,** select the green BAS Multiplier that best describes the given home's climate zone, height above grade (if the basement is half out of the ground, count it as 1/2 story), and wind shielding. Multiply the home's volume by this number.

Figure 1.01.3

# of Stories Above Grade →		1	1.5	2	2.5	3
Zone 1	Shielded from Wind	.1085	.0974	.0869	.0822	.0758
	Normal Shielding	.0904	.0817	.0723	.0689	.0636
	Exposed to Wind	.0817	.0735	.0653	.0618	.0572
Zone 2	Shielded from Wind	.1295	.1167	.1038	.0986	.0904
	Normal Shielding	.1079	.0974	.0863	.0823	.0758
	Exposed to Wind	.0974	.0875	.0776	.0741	.0683
Zone 3	Shielded from Wind	.1505	.1353	.1202	.1143	.1056
	Normal Shielding	.1254	.1132	.1003	.0951	.0881
	Exposed to Wind	.1132	.1015	.0904	.0858	.0788
Zone 4	Shielded from Wind	.1715	.1546	.1371	.1301	.1202
	Normal Shielding	.1429	.1289	.1143	.1085	.1003
	Exposed to Wind	.1289	.1155	.1027	.0980	.0898

Example:

Home Volume: 2500 ft^2 home plus 800 ft^2 basement = 3300 ft^2 total floor area

3300 ft^2 x 8 ft ceilings throughout = 26,400 ft^3

BAS Multiplier: Climate zone 3, 2.5 stories above grade, normal wind shielding =

.0951 Multiplier

Maximum CFM_{50} = 26400 x .0951 = 2511 CFM_{50}

Video demonstration available at www.BuildingPerformanceWorkshop.com/videos

PREPARATION

Close and lock all exterior doors and windows. ***TIP:** *On the first window, verify which way the latch points in 'locked' position, then visually check all other latches in the home.*

Open all the interior doors, including closet doors. You should already have determined which rooms are part of the envelope (*see A.01*).

Fill any dry water traps, and turn all HVAC off (*see A.02*), including exhaust fans.

For fireplaces, ensure that the fire is out, the flue is closed, and the ashes are prevented from becoming airborne. ***TIP**: Use wet newspaper to cover the ashes, and don't forget to take your disgusting ashy wet newspaper with you when you leave.*

Check for asbestos on the envelope, including asbestos fireplace inserts, asbestos insulation on ductwork in exterior spaces, and vermiculite on the attic floor under other insulation (if any or these is found, use 1.01-B Pressurization Method for the test). ***HEADS UP:** *If you even suspect asbestos, play it safe and use 1.01-B. The only way to truly identify asbestos in materials is to test it in a lab, so if it's old and it's not glass or wood, it may contain asbestos.*

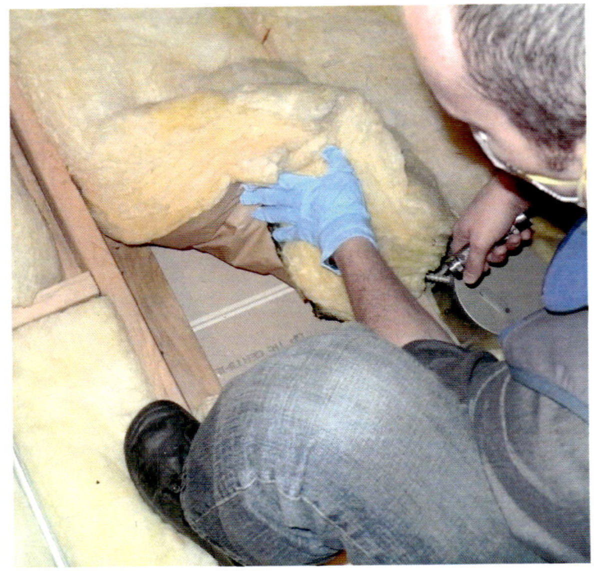

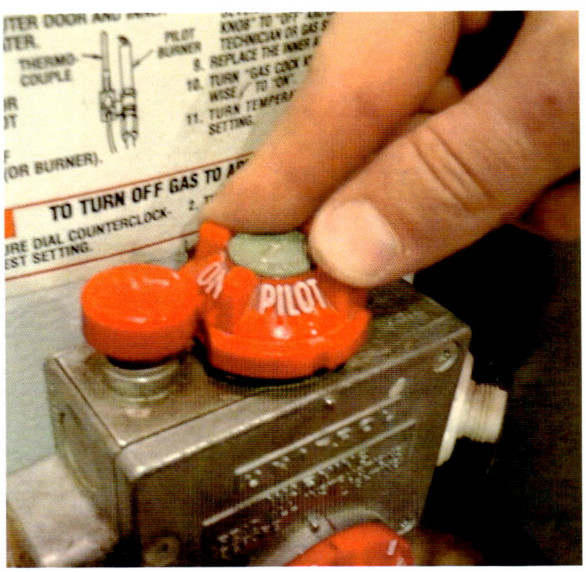

Combustion appliances piloted (in the case of water heaters) or off (using service disconnect or thermostat, in the case of furnaces and boilers) (*See A.02 for instructions*).

Determine which door will be used for the fan. Use an exterior door in a large common area of the home, which opens to an unobstructed space outdoors.

TESTING

Lay the 4 outer pieces of the frame on the floor with the knobs and cams facing you. Connect the 4 pieces so the knobs and cams are showing in 2 corners across from each other. ***TIP:** *If you are inside the frame whenever it's on the floor, your job is easier.*

Fit the frame exactly where door normally sits, and tighten the knobs only. Anything adjustable (knobs) should face you. ***TIP:** *Use one hand to exert pressure in both directions on each frame piece while your other hand tightens the knob.*

Place one end of the long pressure hose outside, well away and off-axis from the fan's airflow path, in a dry and sheltered spot. Leave the other end just outside the door.

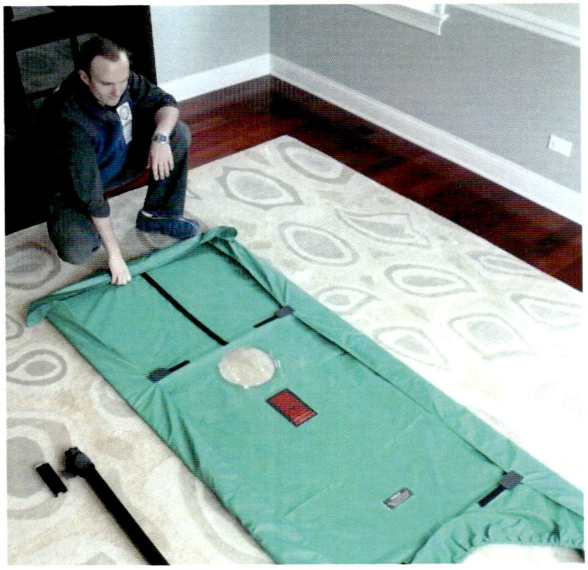

Fit the shroud to the assembled frame and pull it taut. Anything adjustable (like the knobs on the frame and the Velcro on the shroud) should face you.

Install the fully assembled, shrouded frame starting at the bottom, hinged side first, and then lock down the cams. Give the frame a good shake to ensure it won't pop out from the force of the fan.

Install the center brace low enough to hang the fan from, but high enough to provide outward pressure on the frame; taller doors need more lateral pressure at the middle. Tighten the knob, then lock down the cam. ***TIP:** *Ideally, the fan will be suspended off the ground slightly as it hangs from this brace; the fan's weight will add stability to the entire assembly.*

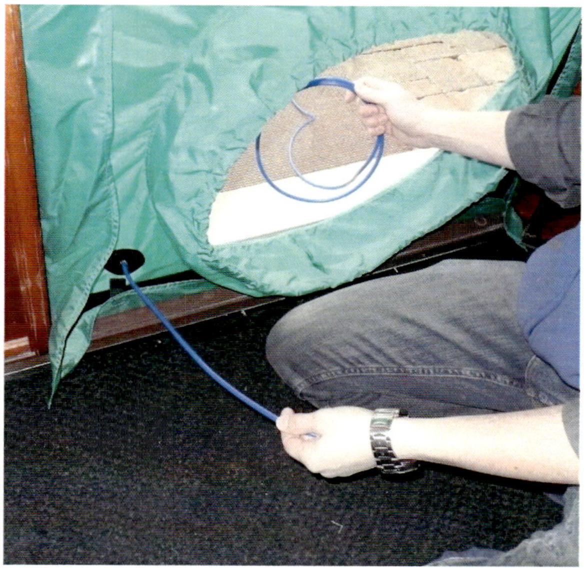

Thread the end of the long pressure hose through one of the hose ports at the bottom of the shroud.

Install the fan in the shroud's hole; attach it to the center brace, and leave the cover of the fan and the flow rings in place. ***TIP:** *Don't fight gravity. Set the fan's base into the shroud's elastic ring first, then slip the shroud over the sides and top.*

Mount the fan control and manometer outside the airflow path, and connect both hoses to the manometer: Channel A measures the pressure in the home with reference to outdoors, and Channel B measures the flow in the fan (*see A.03 for 'Manometer Mechanics'*).

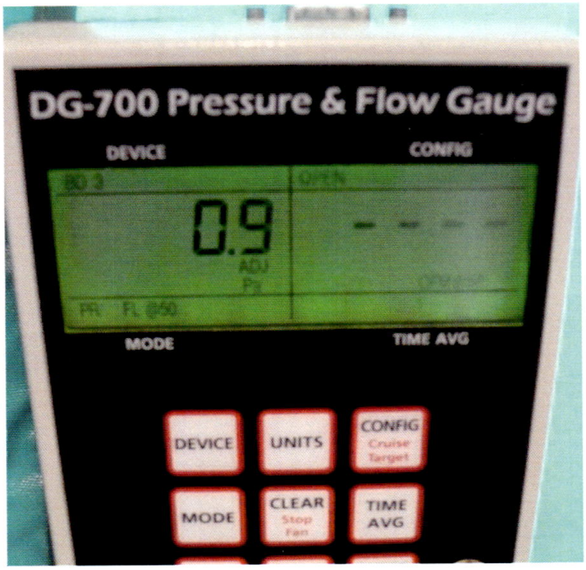

To obtain the CFM50, set the manometer to the correct mode (Pr/Fl@50), device (BD), and ring (Open, A, B, etc), and run the baseline with the fan flow rings in place. ***HEADS UP:** *If your baseline pressure reading is fluctuating because it's a windy day, use a longer baseline sample, and a longer time average for the test.*

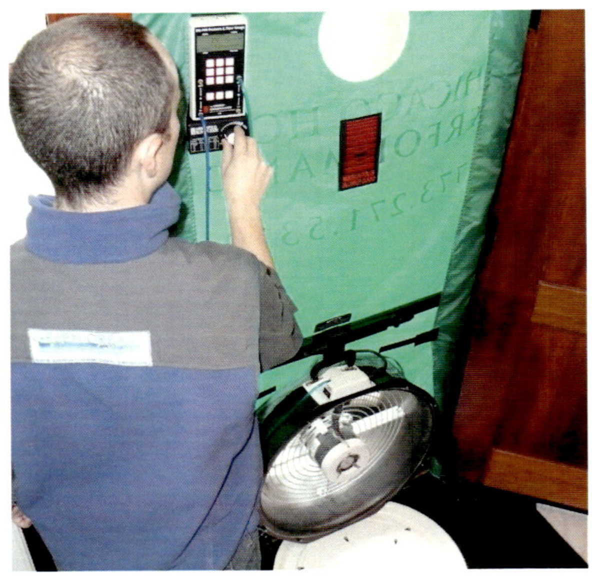

1.01-A Depressurization Method

Remove flow rings and activate the fan, bringing the pressure differential on Channel A to –50 Pa. Record the indicated airflow on Channel B.

1.01-B Pressurization Method

Remove the flow rings, unplug the hose and power cable from fan housing, and uninstall the fan.

Place an additional pressure hose outside, in same location as original reference pressure hose. Bring it in through the shroud's second hole, and attach it to Channel B's reference tap.

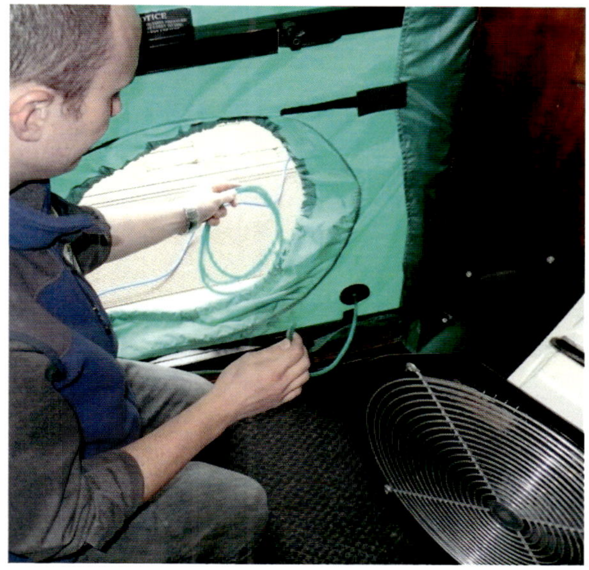

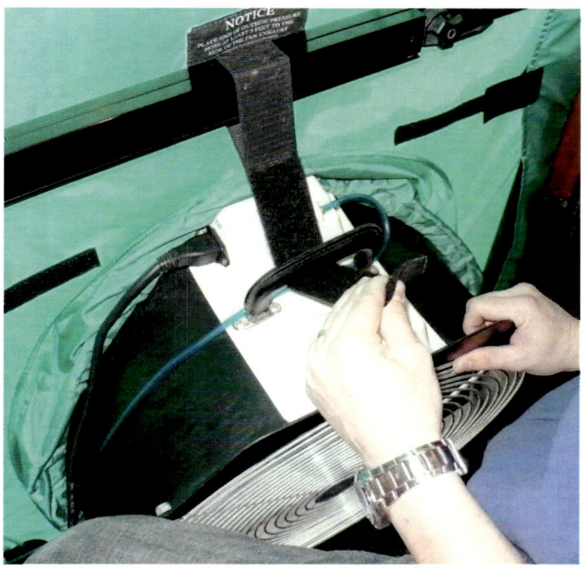

Reinstall the fan so that the exhaust port faces into the home, adjusting the shroud to allow for access to the cable couplings. Reattach the pressure hose and electrical connection.

Double-check the configuration of your manometer. Activate the fan, and bring the pressure differential to +50 Pa. ***HEADS UP:** *If you're testing in an entry vestibule or there are turbulent features in the fan's path to the inside, you should also run a long hose from the Input tap of Channel A into the house, to get the sample location away from the path of the fan.*

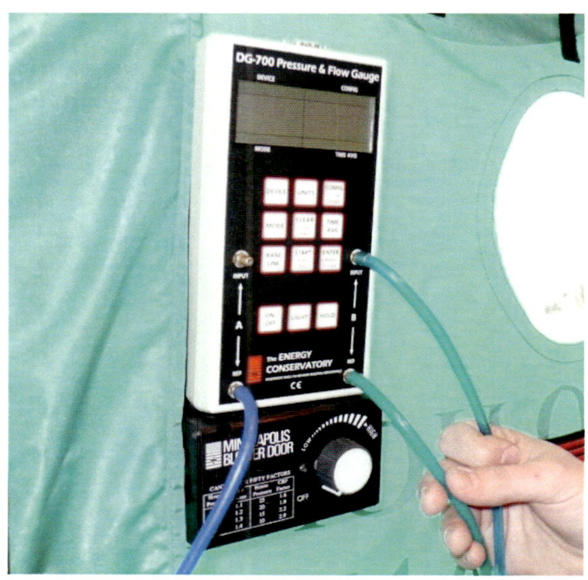

INTERPRETATION

If you get a 'Low' flow reading, add a flow ring to increase the airflow speed through the fan, and configure the manometer to reflect the ring chosen (*see A.03, Manometer Mechanics*).

If you Can't Reach Fifty (CRF), make note of pressure reached. If you are using the "@50" mode on Channel B's flow reading, then the manometer is extrapolating the actual flow through the fan to tell you what it would be at 50 Pa in Channel B's reading. If you are using a simple "Pressure/Flow" mode, then use the Can't Reach Fifty factor from your device manual to extrapolate the flow at 50 Pa.

To help your client better understand the meaning of this test, there are a few different ways to illustrate the results. First, help your client understand the bottom line by calculating how often all the air in the home is completely replaced with outdoor air. You have a CFM_{50} reading or an ACH_{50} rate, which is under blower door conditions, or 'fan-time'. Your client does not live in this dimension; they want to know about 'real-time', which is what happens under natural conditions.

The key to translating between 'fan-time' and 'real-time' is the N Factor. To calculate the N for any given home, let's revisit the Climate Zone map from **Figure 1.01.2**:

Figure 1.01.2

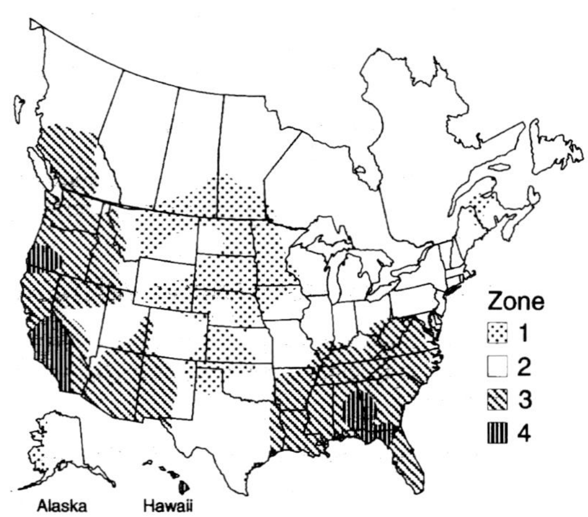

Next, using **Figure 1.01.4,** select the green number that best describes the given home's climate zone, height above grade (if the basement is half out of the ground, count it as 1/2 story), and wind shielding; this is the N Factor:

Figure 1.01.4

# of Stories Above Grade ➔		1	1.5	2	2.5	3
Zone 1	Shielded from Wind	18.6	16.7	14.9	14.1	13.0
	Normal Shielding	15.5	14.0	12.4	11.8	10.9
	Exposed to Wind	14.0	12.6	11.2	10.6	9.8
Zone 2	Shielded from Wind	22.2	20.0	17.8	16.9	15.5
	Normal Shielding	18.5	16.7	14.8	14.1	13.0
	Exposed to Wind	16.7	15.0	13.3	12.7	11.7
Zone 3	Shielded from Wind	25.8	23.2	20.6	19.6	18.1
	Normal Shielding	21.5	19.4	17.2	16.3	15.1
	Exposed to Wind	19.4	17.4	15.5	14.7	13.5
Zone 4	Shielded from Wind	29.4	26.5	23.5	22.3	20.6
	Normal Shielding	24.5	22.1	19.6	18.6	17.2
	Exposed to Wind	22.1	19.8	17.6	16.8	15.4

Use **Equation 1.01.1** to convert between real-time and fan-time for any airflow (in CFM) or air change rate (in ACH):

Equation 1.01.1

$$ACH_{Natural} = ACH_{50} \div N$$
$$CFM_{Natural} = CFM_{50} \div N$$

ACH Example:

$ACH_{50} = 5.0$

N Factor: 16.3 (zone 3, 2.5 stories above grade, normal wind shielding)

$$ACH_{Natural} = 5.0 \div 16.3 = \mathbf{.31\ ACH_{Natural}}$$

CFM Example:

Tested $CFM_{50} = 2511$

N Factor: 16.3 (zone 3, 2.5 stories above grade, normal wind shielding)

$$CFM_{Natural} = 2511 \div 16.3 = \mathbf{154\ CFM_{Natural}}$$

To convert between CFM and ACH, use **Equation 1.01.3**:

Equation 1.01.3

$$ACH_{50} = \frac{CFM_{50} \times 60}{Volume}$$

$$ACH_{Natural} = \frac{CFM_{Natural} \times 60}{Volume}$$

You may also choose to illustrate air leakage in terms of how big the hole would be if you added up all the cumulative gaps, holes, and cracks. There are several different standards for doing this.

The Equivalent Orifice Leakage Area (EOLA) is defined as the area of a sharp-edged hole that would leak at the same flow rate as the estimated leakage rate (CFM_{50}), when the hole is subjected to the target test pressure. The EOLA can be calculated using **Equation 1.01.4.**

Equation 1.01.4

$$EOLA\ in^2 = CFM_{50} \times 0.133$$

Example:

CFM_{50} = 2800 CFM at 50 Pa

$$EOLA = 2800 \times 0.133 = \mathbf{372.4\ in^2}$$

The Effective Leakage Area (ELA), also called Specific Leakage Area (SLA), is a measure based on a theoretical rounded hole that would leak the cumulative amount of all the holes in the envelope with a 4 Pa pressure difference across the envelope (which is a typical pressure difference under natural conditions). The ELA can be calculated using **Equation 1.01.5.**

Equation 1.01.5 $$\boldsymbol{ELA\ in^2 = CFM_{50} \times 0.055}$$

Example:

CFM_{50} = 2800 CFM at 50 Pa

$$\boldsymbol{ELA = 2800 \times 0.055 = 154\ in^2}$$

The Equivalent Leakage Area (EqLA) is similar to the ELA, except it assumes a sharped edge hole at a 10 Pa pressure difference across the envelope. The EqLA can be calculated using **Equation 1.01.6.** It should be noted that all leakage area calculations are empirically derived in laboratory settings, and are merely estimates that can be used to provide a physical indication of a single hole in the envelope based on the flow data collected. All leakage areas are calculated in square inches.

Equation 1.01.6 $$\boldsymbol{EqLA\ in^2 = CFM_{50} \times 0.1}$$

Example:

CFM_{50} = 2800 CFM at 50 Pa

$$\boldsymbol{EqLA = 2800 \times 0.1 = 280\ in^2}$$

QUIZ Blower Door

1) The local energy code mandates that new construction must have a minimum of 5 ACH_{50}. After blower door testing, you find that the infiltration through the envelope is 4964 CFM at 50 Pa. Knowing that the volume of the home is 49,800 ft^3, is this requirement met? ______________

____

____

2) What is the BAS_{50} of the home in question 1 if it is located in Pittsburg, PA? During your inspection, you have concluded that the home is 3 stories tall and is exposed to wind. ____

____

____

3) Name 3 important safety checks that must be performed prior to testing with the blower door and why each is relevant. ________________________

____

____

____

____

4) Upon inspection of the home, you find that there is a presence of vermiculite in the attic after you have set up your blower door test for depressurization, but you have not yet turned your manometer on. What steps are required in order to safely get a CFM_{50} reading for this home?

____

____

____

____

5) A home in Richmond, VA is found to have an infiltration rate of 3547 CFM_{50}. Upon your inspection, you have concluded that the 1.5 story home is shielded from wind. What is the $CFM_{Natural}$ of this home?

____

____

____

6) If the home in question 5 had a volume of 31,200 ft^3, what are the corresponding ACH_{50} and $ACH_{Naturual}$? ______________________

____

____

Discussion and Critical Thinking:

1) Can you think of why it is important to place the reference hose away from and off-axis of the fan flow? How might this affect test results?

____

____

____

2) Why is it important to increase the time average of your baseline reading if it is windy outside when testing? Is it also important to increase the time average during testing as well? ______________

____

____

3) Why do you think wind shielding and height above grade affect the N-factor? ______________________________________

____

____

____

1.02 Infrared Thermography

Materials Needed:

- infrared camera
- chemical smoke
- surface moisture meter

CONCEPT

Infrared thermography is a visual representation of the heat that is emitted, transmitted, and reflected by a material. It is not an X-ray, nor will it show you the movement of air, except where the air brushes against a material and changes its temperature. In order to perform a meaningful infrared diagnostic, you must pay close attention to the temperature outside, the position of the sun, and the variation of both of these over the past several hours.

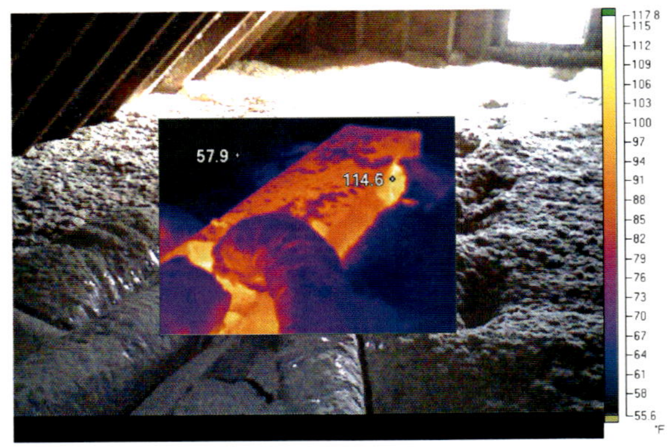

According to standard ASTM C 1060-90, when scanning for conductive heat flow (insulation deficiency or discontinuity) you officially need an 18°F temperature difference between inside and outdoors for 4 hours prior to your inspection, though it is possible to get usable readings with less. If scanning for air leakage, standard ASTM E 1186-03 calls for 9°F temperature difference at the time of testing, but the new RESNET standard states you only need 3°F. You can produce this temperature difference using the home's heating or cooling system if the weather does not cooperate. Consider that buffer spaces (attics, crawlspaces, garages) may be at a higher or lower temperature

Best Practice Tip: Don't look only at the pretty colors. Those are for your client. You care more about the temperatures displayed in the range depicted alongside the image.

than either the home or outdoors; know what that temperature is. Infrared thermography is a non-destructive process, and therefore it can be challenging to reach conclusions with absolute certainty. If in doubt, always put it this way: "It appears to be that..." the wall is uninsulated, or perfectly insulated. Ultimately, the only way to be 100% certain is to drill a hole or cut a hatch and see for yourself.

PREPARATION

Note the temperature outdoors and the position of the sun over the past several hours. As far in advance as possible before your IR scan, move furniture and wall hangings away from any building surfaces you will want to accurately analyze.

TESTING PRIOR TO THE BLOWER DOOR:

Turn on the camera and adjust the settings to manual temperature calibration (this is to remove the accidental re-calibrating effect of scanning a piece of the sky during your exterior inspection). The low and high temperatures should be equal to outside air temperature and 15° closer to inside air temperature, respectively.

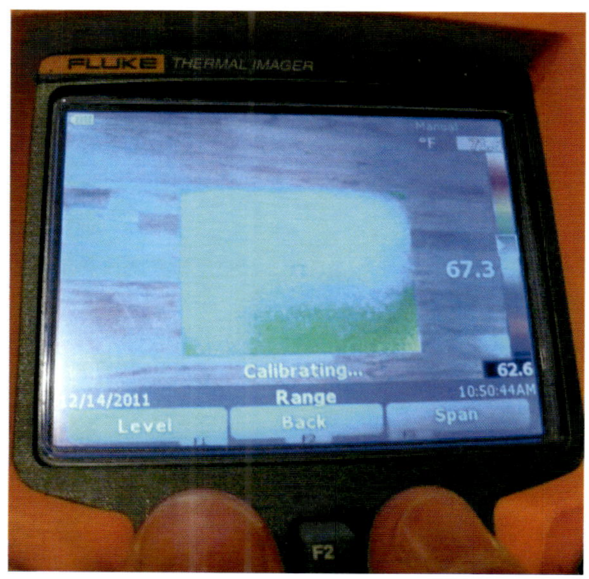

Inspect the home from outside, taking note of any temperature variations, especially those that approach the temperature of the living space or HVAC. These variations indicate weaknesses in the envelope. In a home where heat flow and airflow are well controlled, your infrared camera should show you very boring images, with no variation at all. **HEADS UP:** *Do* ***not*** *pay attention to the glass of windows with your IR camera. They will show a much larger percentage of reflection than other building materials, so if you're shooting up at a second story window, the temperature you're seeing is heavily influenced by the temperature of the sky (and of outer space beyond it, which is pretty cold).*

Inside, inspect the envelope from the interior (*for more on knowing where the envelope actually is, refer to A.01*). Feel free to set your camera to 'Automatic' Range calibration at this point. Look for insulation deficiency or discontinuity, and look for moisture. Don't bother recording images of air leakage; they will look much clearer and more dramatic under blower door depressurization. ***TIP:** *All corners or joints between materials will typically show heat flow, so don't worry about them unless they also show air leakage during the blower door test.*

Beware of readings influenced by the radiance of the sun. Always keep track of its position in the sky relative to your location in the home. What looks like insulation settling might just be the eaves of the roof casting shadows.

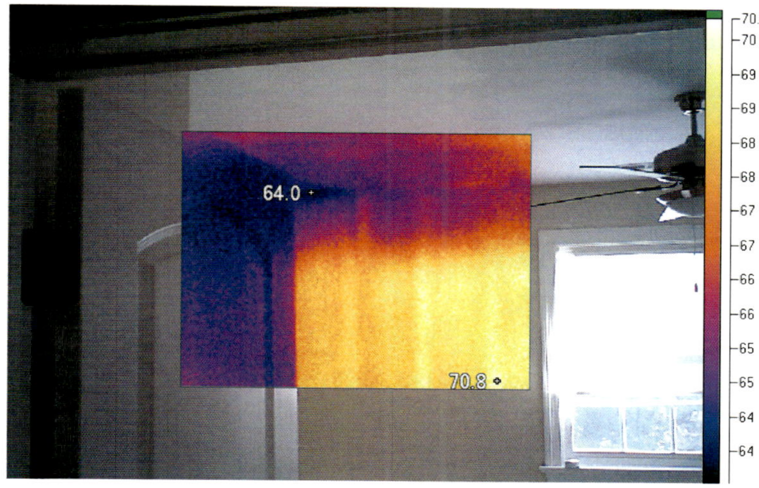

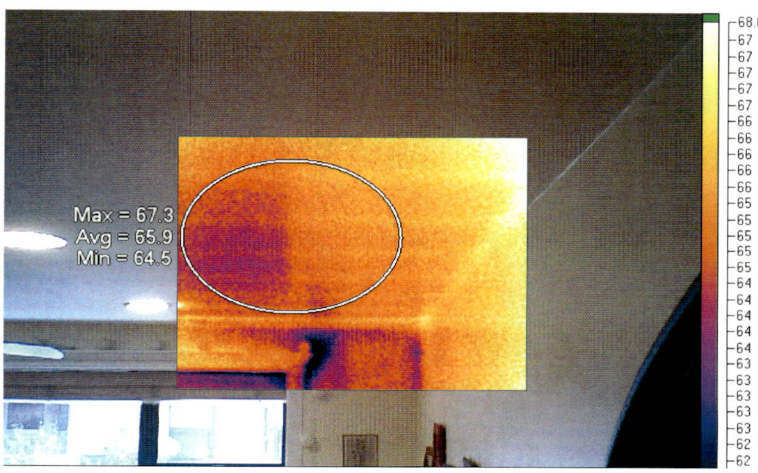

Beware of readings influenced by reflections; almost every surface will reflect for your infrared camera. Before you take an anomalous image seriously, move to your right or left and see if the anomaly moves with you.

TESTING DURING THE BLOWER DOOR DEPRESSURIZATION TEST:

Begin on the most seasonally depressurized level of the home (due to stack effect, it's the lowest floor in winter, the highest floor in summer) and scan the entire envelope, working your way toward the most pressurized level.

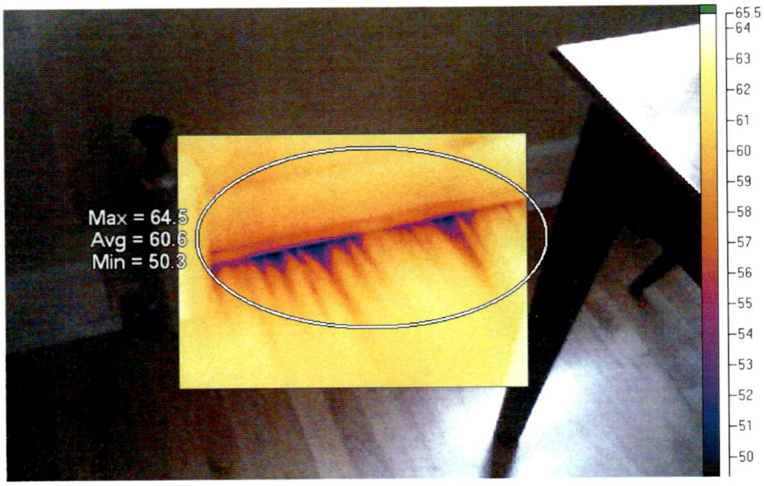

When pinpointing air leakage through the envelope (at windows, doors, penetrations, etc.), verify your infrared reading using chemical smoke, to confirm that you've found a significant leak. Sometimes air leakage looks significant, but might actually be a very small amount of air moving very fast at the perfect angle to the building materials; this is not something you should spend a lot of time worrying about. ***TIP:** *Always consider the usefulness of the images you record; for instance, if you fill a report with tons of images of baseboard leakage, as in this photo, your client might spend tons of time and energy caulking all the baseboards, while the real source of the air leakage is likely the band joist behind and below it.*

Look for readings that approach outdoor air temperatures, or buffer space temperatures (for example, on a 70°F day, a home with a black roof will have an attic temperature of over 100°F). ***TIP:** *Use zonal pressure testing first (see 1.03) to determine which rooms will reveal significant air leakage; that way you don't waste time scanning for air leakage in zones that have none.*

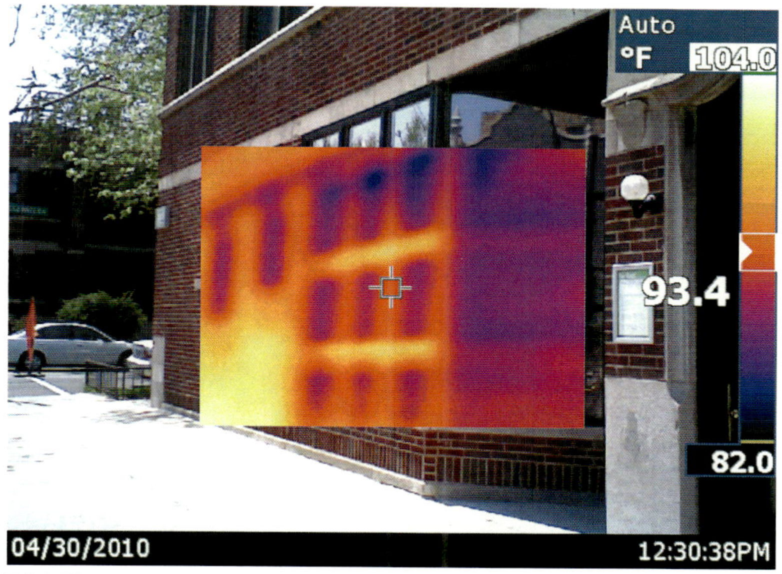

On cameras with both infrared and digital (visual spectrum) images, intentionally unfocus the infrared for shots meant to be visible-spectrum only (for ease of report building later). ***TIP:** *Photograph hand signals at arrival on each floor, for ease of report building later. For example, a picture of your fist for BSMT, 1 finger for first floor, etc.*

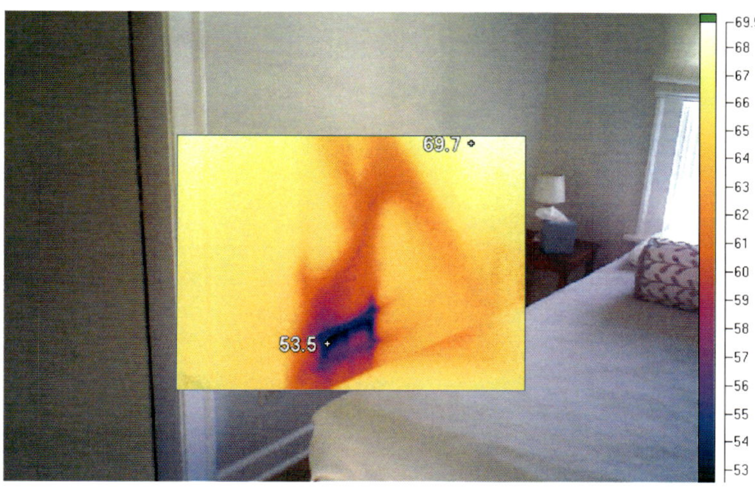

Your camera may show you evidence of duct leakage: in winter, if the registers are nearing outdoor air or buffer space temperatures, and you know that just a few minutes ago those ducts were blowing hot air from the furnace, you can deduce that the registers are being cooled by outside air making its way into the ducts. Always verify with a pressure pan test (*see 1.04*) and visual inspection.

INTERPRETATION

Liquid Water Leakage is sharp, uniform in color, falling under gravity, and verifiable with a surface moisture meter.

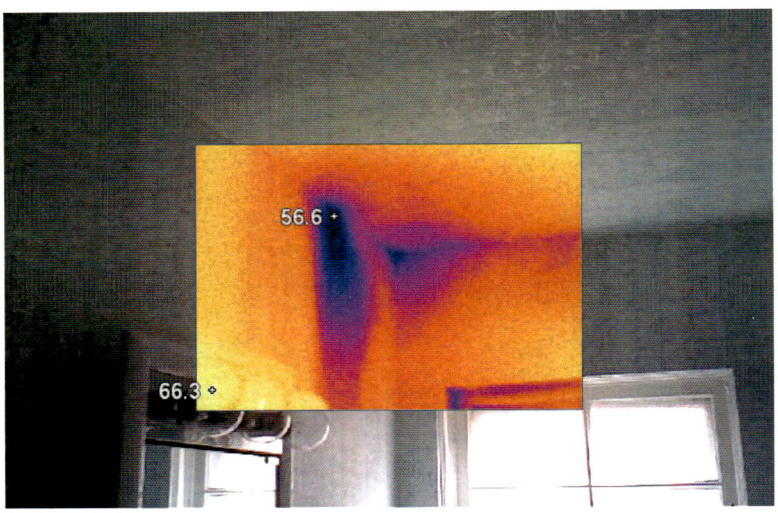

Vapor Condensation is sharp, graded in color, verifiable with a surface moisture meter, and often indicates an air bypass.

Insulation Deficiency is soft and bruise-like, and may be difficult to focus on with your camera. It is verifiable through visual inspection.

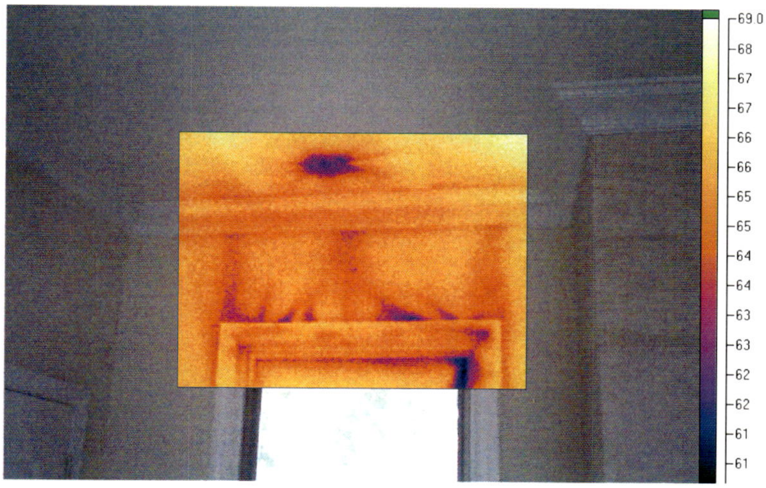

Air Leakage is sharp, finger-like, and verifiable with chemical smoke. ***TIP:** *Don't look for air leakage until the blower door runs. Because of stack effect, you won't see air leakage in any pressurized levels of the home, since drafts there would be going out through the envelope, not coming into it.*

Insulated or Uninsulated? If you see that all the framing members look uniform and all the cavities look uniform, then the wall must either be perfectly insulated or perfectly uninsulated. Heat always moves from high to low temperature, so simply

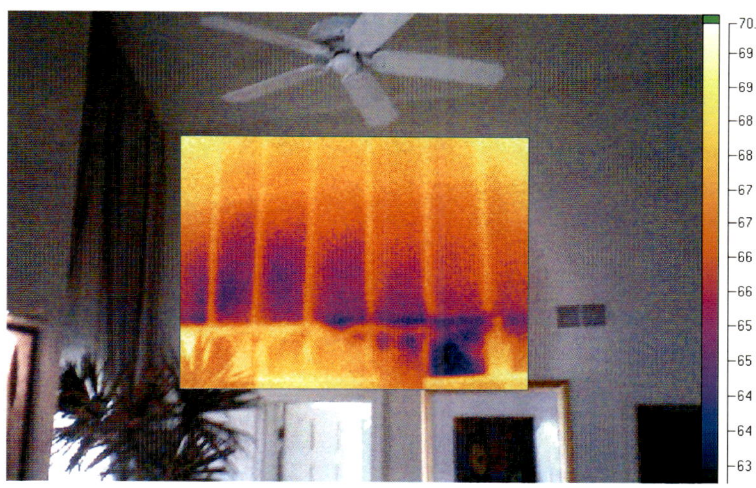

be aware of whether heat is moving into the home (summer) or out of it (winter), and determine whether it's flowing faster through the framing or the cavities. Where heat is moving faster, there must be less insulation to control its flow.

QUIZ Infrared Thermography

1) The American Society for testing and Material has set forth guidelines for different types of infrared inspections. What are they? ____________________

____

____

2) Describe a few instances where wall assemblies/components or ambient condition can falsify an image and how to avoid these situations. __________

____

____

3) Why is it important to know the temperature of 'buffer zones' in a home when scanning with an infrared camera during blower door testing? _______

____

____

4) What are some simple tricks that the thermographer can use to help indicate the location where an image was taken? ________________________

____

____

5) How can an infrared camera show indications of duct leakage? Can you think of a way to confirm this? ____

____

____

6) Describe the key signatures of liquid water leakage, vapor condensation, insulation deficiencies, and air leakage. ______________________

____

____

____

Discussion and Critical Thinking:

1) What is the benefit of beginning your infrared scan (during blower door testing) in the most seasonally depressurized level of the home?

____

____

____

____

2) As a time saver, how can Zonal Pressure Testing (Section 1.03) assist the infrared scanner when he/she is looking for air leakage with the blower door? ______________________

____

____

____

3) Which direction does heat flow? Knowing that infrared cameras only detect surface temperature, describe what an uninsulated wall would look like in wintertime and summertime if scanning from the interior. How about an insulated wall? ______________________

____

____

____

____

1.03 Zonal Pressure Testing

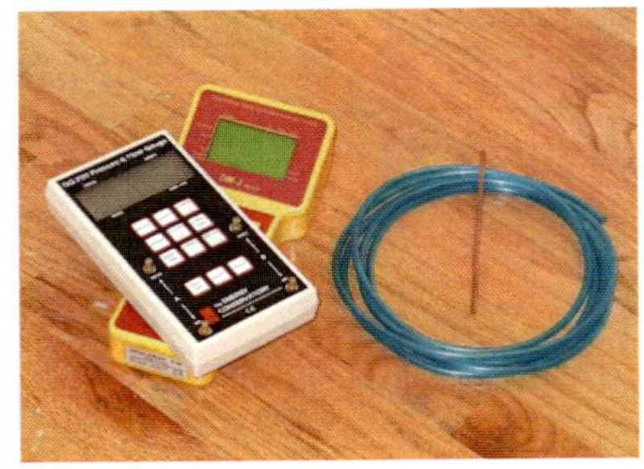

Materials:

- digital manometer
- short pressure hose
- 5/32" metal probe

Video demonstration available at www.BuildingPerformanceWorkshop.com/videos

CONCEPT

The zonal pressure test is a comparative measure between zones in the home that show either:

A. opportunity for improvement in envelope air sealing (**1.03-A Envelope Air Leakage Method**), or

B. pressurization/depressurization from HVAC (**1.03-B HVAC Pressure Effects Method**)

Best Practice Tip: For testing Envelope Air Leakage, using the automated blower door setting ('cruise control') is always best. The home remains at the same pressure, despite all the doors you'll be closing during this process; hence, your data is as accurate as possible. YOU WILL NEED A SECOND MANOMETER TO DO IT THIS WAY.

Pressure testing is quantitative by nature; it is based in numbers. A zone's connection to the outdoors can be represented as a percentage. For example, during a blower door test at 50 Pa, a zone shows a pressure differential of 5 Pa; this means the zone is 10% connected to outdoors. A zonal pressure reading of 45 Pa under the same conditions means the zone is 90% connected to the outdoors.

It's important to understand the qualitative nature of zonal pressures, as well. For example, if the second floor bathroom and the entire basement both show readings of 15 Pa during the blower door test, you'd miss out on the power of this test if you failed to notice that the size of the bathroom and basement are wildly different. The bathroom now represents a much bigger opportunity for improvement, due to your qualitative analysis.

A comparative analysis of test readings will help you prioritize recommendations. If, under natural conditions, the central HVAC system causes a bedroom to become pressurized to +5 Pa, and the bedroom across the hall reads +15 Pa, then both should be fixed, but the room with the higher reading should be fixed first.

PREPARATION

Set up your manometer to measure pressure in a zone (room, crawlspace, attic, garage, etc.), with reference to the home (*for more on this subject, see A.03, 'Manometer Mechanics'*).

1.03-A Envelope Air Leakage Method

Run the blower door according to 1.01 and follow the testing procedure, focusing on rooms of the home and buffer zones.

1.03-B HVAC Pressure Effects Method

Without the blower door running, turn on the air handler according to A.02 and follow the testing procedure, focusing on rooms that have duct registers.

TESTING

With the end of the 'input' hose in the zone, close the door and measure the pressure in the zone with reference to the home. ***TIP:** *For tight fitting doors, attach a metal probe to the end of the pressure hose to keep from pinching the hose.*

INTERPRETATION OF ENVELOPE AIR LEAKAGE:

The pressure the home has reached using the blower door is the maximum possible reading, and now represents a 100% connection to outdoors. For example, if you conduct your envelope testing at +/–25 Pa to minimize the loss of conditioned air (after obtaining the CFM @ 50 Pa number, of course), then a zonal pressure reading of 25 will mean the zone is completely connected with outdoors.

If you always reference the home, then a reading of 0 Pa will mean the zone is completely connected with the home.

A reading that falls between the blower door induced house pressure and 0 Pa means the zone has connections with both the home and outdoors, and represents an opportunity for improvement through air sealing. The closer it is to 0, the easier it would be to include it in the envelope. The closer it is to the blower door induced pressure, the easier it would be to exclude it from the envelope.

INTERPRETATION OF HVAC PRESSURE EFFECTS:

If the pressure in the zone is positive, there is more air being supplied to the zone than returned from it. If the pressure in the zone is negative, there is more air being returned from it than supplied to it.

If the pressure in the zone is higher than ±3, the zone is unable to have a functional relationship with the air handler, and will thus become uncomfortable. Pressure relief between the zone and the main body of the home is recommended in this case, through an undercut door, transfer grille, or jumper duct.

QUIZ Zonal Pressure Testing

1) Describe how this test is both qualitative and quantitative. How would you convey information to a contractor using both methodologies? ____________

____

____

2) The blower door has the home held at -50 Pa. You test the unconditioned, ventilated attic and the manometer is showing a 35 Pa reading. What does this mean? What reading would you like to see? ____________

____

____

3) If a home has several 'buffer zones' that are confused in regards to whether they are part of the conditioned envelope, or outside the conditioned envelope, how would you use zonal pressure testing to help make recommendations in regards to air sealing? ____________

____

____

4) What are the upper and lower pressure limits for induced HVAC pressures? How can you provide relief to a room that exceeds these limits? ____________

____

5) At -50 Pa, you test the pressure in a basement referencing the interior and it shows a reading of +10 Pa. You then have your technician who is outside (and therefore referencing outside) test the basement through the rear door. What reading would you expect the technician to see? ____________

____

____

Discussion and Critical Thinking:

1) Why would it be beneficial to have your blower door set to automated setting ('cruise control') when testing for envelope leakage around the home? ____________________

____

____

2) You want to pressure test a master suite that has a bathroom, wash closet, and separate closets for the occupants sharing the room. How would you approach testing the room if each sub-zone has closable doors? ____________________

____

____

3) Discuss the relationship of a 'buffer zone' connection to the house. If there is ductwork running through a 'buffer zone' that has connection to house and outdoors, how would this affect pressure pan testing results?

____

____

____

____

1.04 Pressure Pan Testing

Materials Needed:

- digital manometer
- pressure pan
- pressure hose

CONCEPT

The pressure pan test is a zonal pressure test for a building assembly that does not have a door. You use the solid barrier of the pan to create a temporary 'door' to this 'zone'. Therefore, please read Zonal Pressures 1.03 before tackling this test.

TESTING

Run the blower door test according to 1.01, and leave it running. Set up your manometer to measure the pressure in the pan with reference to the home.

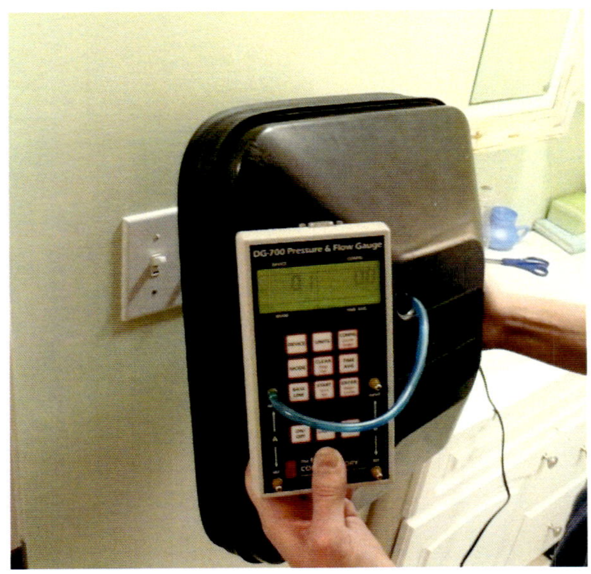

Completely cover a duct register, exhaust fan, recessed light, outlet, or light switch with the pressure pan. Use a steady hand; if the pan moves, the pressure will fluctuate. Read the pressure in the pan with reference to the home.

INTERPRETATION

For exhaust fans: a reading that matches the blower door pressure indicates that the fan is ducted directly to outdoors (this is good). Readings below that mean the fan housing, ducting, or terminal leak air to the house.

For outlets and light switches: the zone being tested is the wall cavity. Interpret readings according to zonal pressure testing protocol (*see 1.03-A*).

For ductwork with significant portions located in unconditioned space, with the blower door running, follow these criteria:

- If all readings are less than 1.5 Pa (3%), and not more than two readings are over 1.0 Pa (2%), then the duct system is considered tight and no further testing or duct sealing is recommended.
- If three or more readings are greater than 2.0 Pa (4%), duct sealing is recommended, prioritizing duct runs with the highest readings first.
- If the readings don't meet either of the above descriptions, then duct leakage testing, both 'total' and 'to outside', (*see 2.09 and 2.10*) is recommended.

Best Practice Tip: Always perform zonal pressure testing on buffer spaces containing ductwork first, because in all likelihood they are not actually 100% connected to outdoors, and so the pressure pan readings will be undervalued. If the crawlspace zonal pressure reads a 50% connection to outdoors when the blower door is running, then the pressure pan maximum reading for ducts in that crawlspace should be much smaller than the criteria listed.

For ductwork mostly inside the envelope: results that indicate leakage as described above represent an even *bigger* opportunity for improvement, since the ducts are not near the envelope, and therefore should not be leaking to outside at all.

QUIZ Pressure Pan Testing

1) Describe some common components and assemblies that should be evaluated when pressure pan testing. ______________________________

2) When covering duct registers, what range or readings is acceptable when the house is pressurized to -50 Pa? When is further testing for duct leakage a good idea? ______________________________

3) If you zonal pressure test an attic and find that 90% connected to outdoors, what is the highest pressure reading you will see on the manometer when testing a can light that leads into the attic? The home was tested at -50 Pa. ______________________________

Discussion and Critical Thinking:

1) How is pressure pan testing similar to zonal pressure testing? ______________________________

2) Describe the benefits of using a pressure pan. ______________________________

1.05 Delta Q

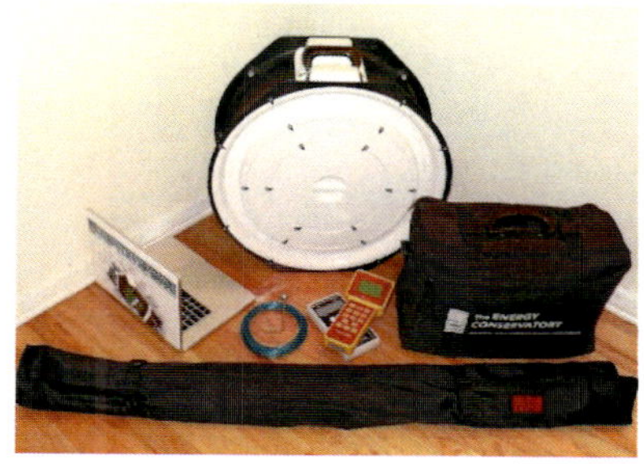

Materials Needed:

- blower door (complete kit)
- functional air handler attached to ductwork
- manometer
- static pressure probe or pitot tube
- pressure hoses
- computer with internet access to ΔQ software calculator available at: *www.BuildingPerformanceWorkshop.com/deltaq*

CONCEPT

Best Practice Tip: Don't attempt this test on a breezy day. Wind has a profound effect on the accuracy of the test results. Don't try it drunk, either, it's no fun at all.

The Delta Q protocol is technically a duct leakage test, but it utilizes the envelope and standardized envelope testing, so I've located it in this section of the book.

Delta Q (Δ, the symbol for change, and Q, the symbol for flow) was developed by Lawrence Berkeley National Laboratory as a timesaving alternative to duct leakage testing. It also has the benefit of being a more realistic test of duct leakage than duct leakage testing (*see 2.09 and 2.10: those tests are actually more accurately described as duct tightness tests, since they do not account for changes in static pressure throughout the duct system*). Delta Q is generally not accepted by formal weatherization or building efficiency programs because of the high potential for user error. But it is fascinating, and may be a useful diagnostic for demonstrating actual duct leakage under normal operating conditions, so I've included it here for your reference.

One criticism of this test protocol is the amount of calculation that happens 'behind the curtain'. A computer is generally required to perform the calculations that reveal the air leakage results; therefore, you have no means of checking or arguing with the results. I've included it here to expose you to a simple application of this obscure test, but if you were to do it as accurately as possible, you should understand the deeper science and formulas behind the test by reading ASTM E-1554 and the publications from Lawrence Berkeley National Lab. Also, the more accurate version of this test uses a ramping measurement of the pressures, and not the pressure stations described below; for this ramping, you must control the blower door using a computer, and record the air flow results using software designed by the blower door manufacturer. Energy Conservatory has refined a Delta Q calculator using this more complex testing style, which they will share with you if you ask nicely.

PREPARATION

Prepare for the blower door test in depressurization mode according to 1.01-A, up to and including baselining the manometer.

TESTING

With the air handler fan on, measure the external static pressures in Pascals (Pa) in both the supply and return plenums according to 2.02. Note that both pressures are recorded as positive numbers for use in the analysis (i.e., the return pressure is NOT negative). Leave the air handler running.

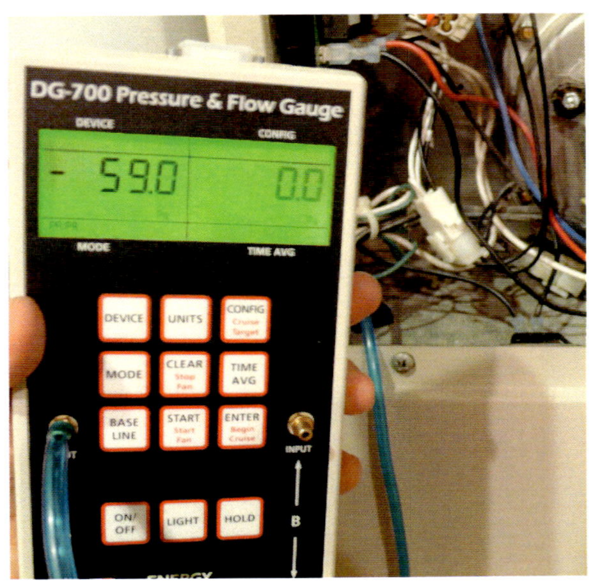

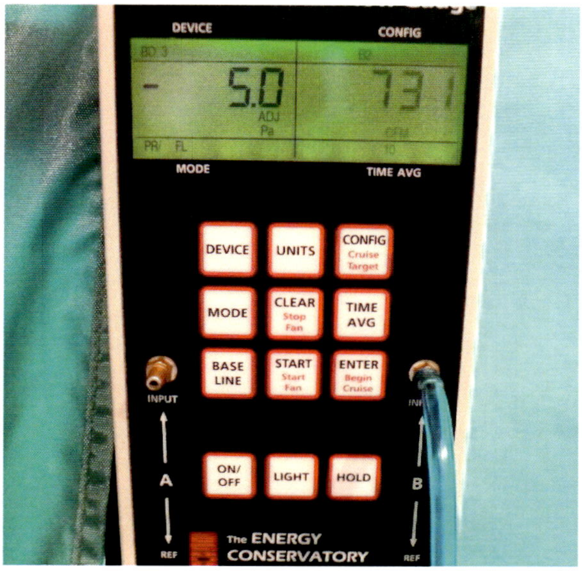

A. Turn on the blower door in depressurization mode (*see 1.01-A*) and bring the home to –5 Pa with reference to (WRT) outdoors. Record the flow through the fan at this target pressure (called a 'pressure station'). Only record pressure and flow readings when the pressure reading is within 1.0 Pa of the pressure station. Use a 10 second time average, and take 3 individual pressure and flow readings at each pressure station, to be averaged for use in the calculation procedure. ***TIP:** *Don't waste time; use the shortest time average available on your manometer to hone in on the pressure station, then switch to a 10 second time average.*

B. Repeat step A, but with the home's pressure difference incremented by –5 Pa each time until the envelope pressure difference is –25 Pa. At each pressure station the pressure difference must be within 1.0 Pa of the pressure station. ***TIP:** *You will probably need to change flow rings on the fan at some point between pressure stations. When in pressurization mode, this means you have to go outside to add the rings. Don't forget to calibrate your manometer to the new flow ring settings when you do this.*

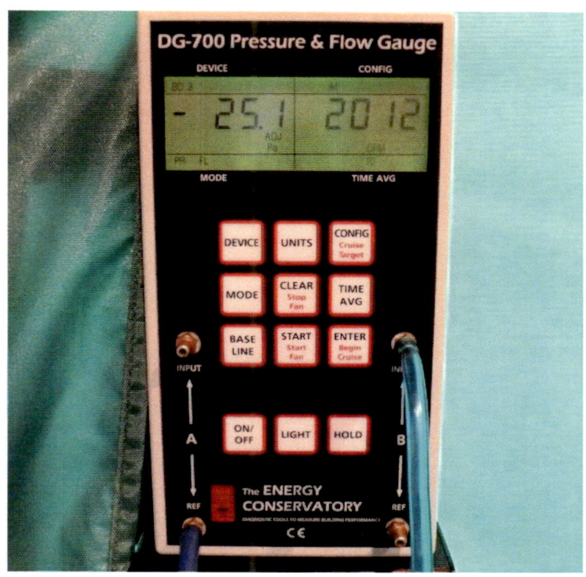

C. Turn off the air handler fan and repeat steps A and B, recording the fan's airflow and exact pressure at each pressure station.

Configure the blower door for pressurization testing (see *1.01-B*). Turn the air handler on and repeat steps A, B, and C, with the house pressurized. The pressure stations at which you're measuring blower door flow are +5, +10, +15, +20, and +25 Pa.

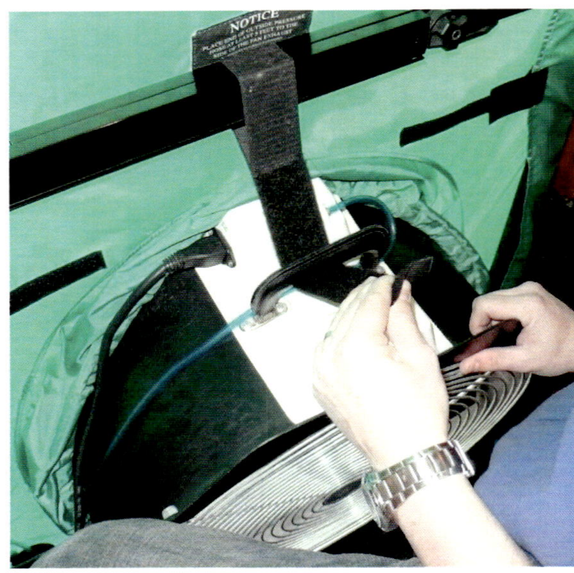

Enter the test data into the Building Performance Workshop Delta Q Calculator at:

www.BuildingPerformanceWorkshop.com/deltaq

QUIZ Delta Q

1) Discuss the benefit of 'honing in' to the pressure station using a short time average, then switching to a 10 second time average. ____________________

____

____

____

2) Why do you think you will need to change flow rings throughout the test?

____

____

____

Discussion and Critical Thinking:

1) Why do you think the 'ramping' method of this test would provide for more accurate results? ____________________

____

____

____

2) What do you think happens when the pressure in the home exceeds that of the ductwork? ____________________

____

____

3) Would you expect the static pressure in the supply and return plenums to remain the same throughout the entire test? Why or why not? ________

____

____

____

2 HVAC DIAGNOSTICS

2.01 System Airflow Testing

CONCEPT

Proper total system airflow is a key part of a high performance forced air HVAC system. There are five methods demonstrated here, in order of accuracy and convenience. The Pitot tube is the gold standard against which all other airflow testing devices are calibrated, so it comes first. Of the other methods, Pressure Matching with a Powered Flow Hood is generally regarded as the most accurate, though it's complicated to set up, which is why it's further down the list.

Causes of incorrect system airflow can include dirty filters, fans, or A/C coils; blockage or closing of dampers, incorrect fan selection or speed settings, or even the ductwork design and layout. You should make sure the system is as clean and unblocked as possible before testing (unless you're trying to prove a point about cleanliness to your client).

Standard system airflows should be roughly 400 CFM per ton of cooling. Heating airflows may be lower or higher, based on the design of the system; therefore, you should test in cooling mode if possible, since you know what the flow should be. If it's under 60°F, however, don't run the air conditioner's compressor; you'll need to trick the air handler into running by itself on high speed, according to A.02.

WARNING: Running an air conditioner's compressor when the outdoor temperature is under 60°F can damage the system. See A.02 if you have doubts about how to avoid this.

Before running any of the tests, refer to the A/C condenser unit label to know how many CFM you're expecting to find (*see A.02, 'HVAC Documentation and Controls'*). Approximate airflows are found in **Figure 2.01.1**.

Tons of A/C	CFM
1.0	400
1.5	600
2.0	800
2.5	1000
3.0	1200
3.5	1400
4.0	1600
4.5	1800
5.0	2000

Figure 2.01.1

2.01-A Velocity Pressure Method (Pitot Tube)

Materials Needed:

- Pitot tube (5/16" diameter, 24" insertion length)
- digital manometer
- two short pressure hoses
- probe thermometer
- 12 or 18 volt drill and drill bit (1/3" minimum)

CONCEPT

A Pitot tube incorporates 2 probes in one tool: the outer probe measures static pressure, and the inner probe measures total pressure, which includes both static and velocity pressure, as **Figure 2.01.2** illustrates.

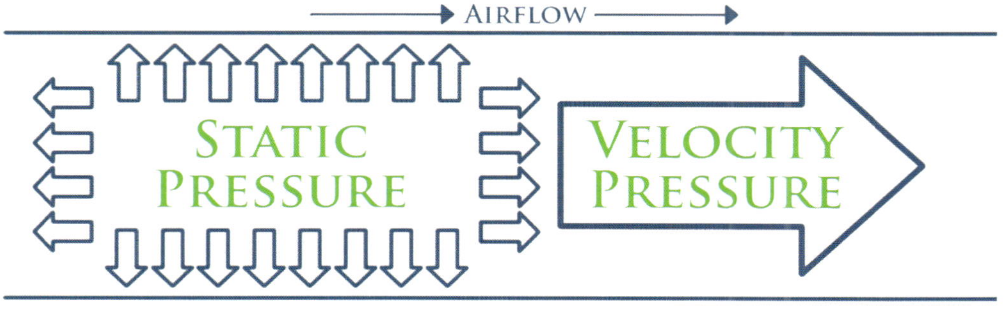

Figure 2.01.2

Once you traverse the duct to establish the velocity pressure (by subtracting the static pressure from the total pressure), you can calculate airflow accurately. An air density correction is necessary, and is included in the 'Interpretation' section.

PREPARATION

Turn off the air handler according to A.02. Drill three equidistant 3/4" holes in the return plenum, in a line perpendicular to the airflow, and away from bends or turbulent features. If the duct system is full of turbulent features, you may need to choose the 'least bad' place to drill the hole. Do your best. ***TIP:** *If you only need a ballpark measurement, you can take a single measurement at the dead center of the plenum, instead.*

Ensure the filter slot is sealed, either with an installed filter slot cover or with tape. This ensures all the airflow will be measurable by the Pitot tube, and none will sneak in through the filter slot.

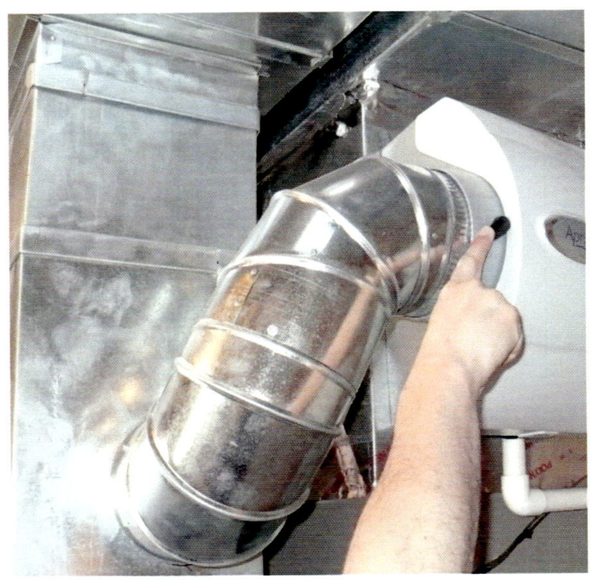

If there is a central humidifier that incorporates a duct from supply plenum to return plenum, make sure the bypass damper is closed. You don't want the temperature of the return air changing during the test; that would also affect the air's density.

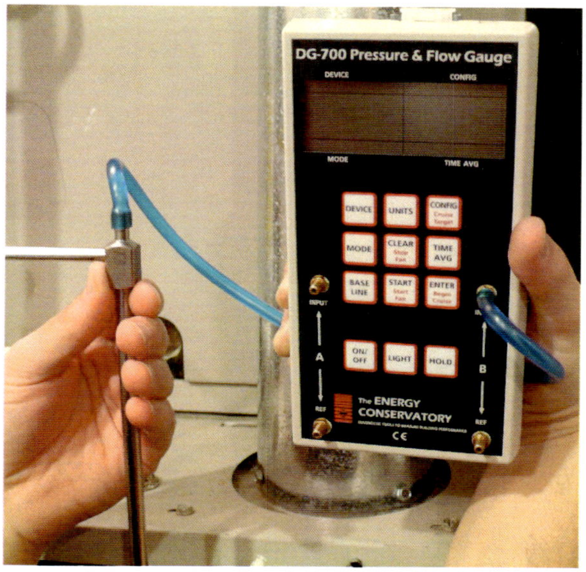

Connect one end of a pressure hose to the Pitot tube's total pressure tap (the smaller diameter), and connect the other end to the Input tap of either channel on the manometer.

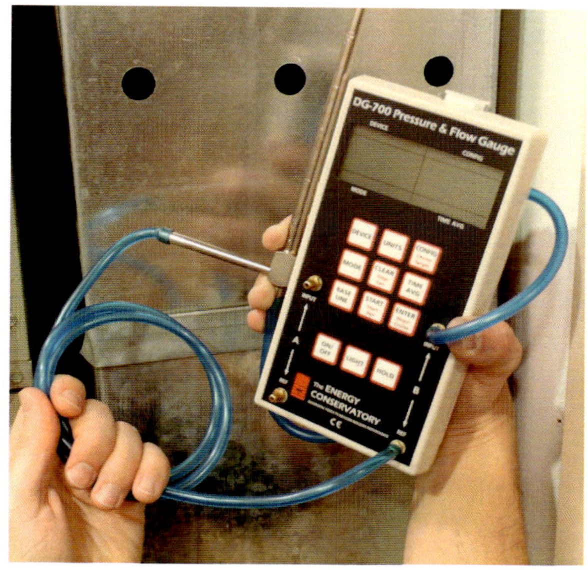

Connect the second pressure hose to the Pitot tube's static pressure tap (the larger diameter), and to the same channel's Reference tap.

Measure the width and depth of the plenum at the point of measurement.

Turn on the manometer, and set it up to measure pressure differential in inches of water column (inWC) (*see A.03*). ***TIP:** Even if your manometer is capable of measuring air velocity on Channel B, measure the pressure instead; you'd still have to correct for air density later (see **Equation 2.01.1**).*

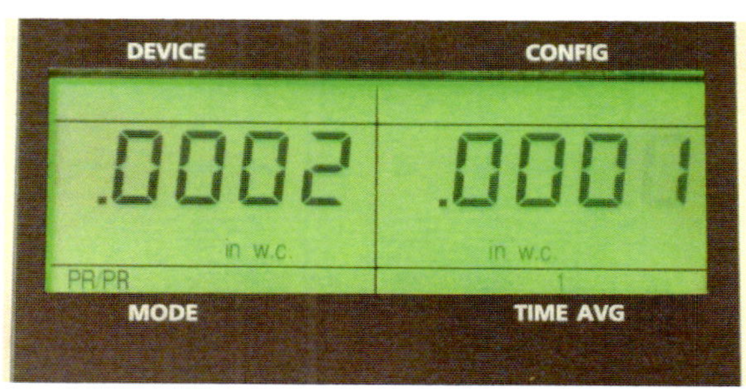

Pointing into the airflow, insert the Pitot tube completely into the first drilled hole in the return plenum. You'll feel the instrument touch the far side of the duct.

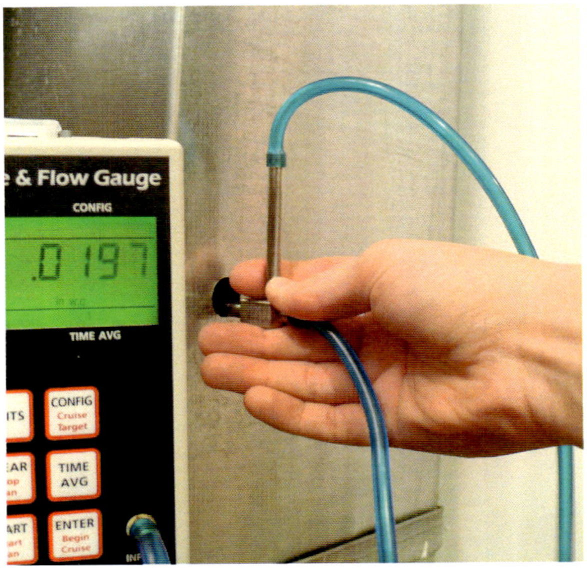

You're about to use a timed average measurement, so just before you begin the traverse from one side of the plenum to the other, set the Time Average on the manometer to 'Long', or as long as it takes you to traverse the duct through each hole, and record the pressure or velocity reading once you complete each traverse. ***HEADS UP:** *Do a practice run, sliding the Pitot tube smoothly from its completely inserted position back toward you until it has crossed the entire duct.*

TESTING

Turn on the air handler in cooling mode, according to A.02. With the Pitot tube pointing continually into the airflow, begin the time average or multi-point average, and traverse the duct by pulling the instrument back out toward you. Measure the velocity pressure in inWC.

At the same time, measure the air temperature in the return plenum.

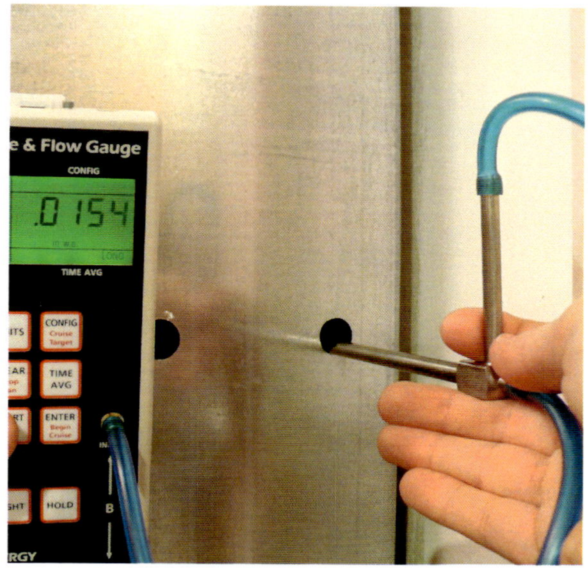

Repeat the process for the second and third holes, and average the measurements across all three traverse lines.

INTERPRETATION

First, calculate the density of the measured air using **Equation 2.01.1**:

Equation 2.01.1 $$d = 1.325 \times \frac{P_B}{T + 460^{\circ}}$$

Where:

d = Density of air [lbs/ft^3]
P_B = Barometric static pressure [inHg, inches of Mercury]
T = Temperature of air in the duct work at testing location [°F]

If you don't know your local barometric pressure, refer to A.07.

Example:

P_B for test home in Denver, at 5000 ft elevation = 24.89

T = Air temperature in the return plenum is 74.4°F

$$d = 1.325 \times \frac{24.89}{74.4 + 460} = .0617\ lbs/ft^3$$

Now calculate the velocity of the tested air using **Equation 2.01.2**:

Equation 2.01.2 $$V = 1096.7 \times \sqrt{\frac{P_v}{d}}$$

Where:

V = Velocity of airflow [ft/min or FPM]
P_v = Velocity pressure [inWC]
d = Density of air from **Equation 2.01.1** [lbs/ft^3]

Example:

P_V = average of .0171inWC

d = .0466 lbs/ft^3

$$V = 1096.7 \times \sqrt{\frac{.0171}{.0617}} = 577.4\ ft/min$$

And finally, calculate the total airflow through the system using **Equation 2.01.3:**

Equation 2.01.3 $$CFM_{Duct\ System} = V \times A_{Duct}$$

Where:

V = Velocity of airflow from **Equation 2.01.2** [ft/min or FPM]

A_{Duct} = Cross sectional area of the duct at the location of pressure measurement [ft^2] *(See A.06 for geometric formula reference sheet)*

Example:

V = 577.4 FPM

A_{Duct} = 10" x 24" = .83' x 2' = 1.66 ft^2

$$CFM_{Duct\ System} = 577.4 \times 1.66 = 958\ CFM$$

2.01-B Flow Plate Method

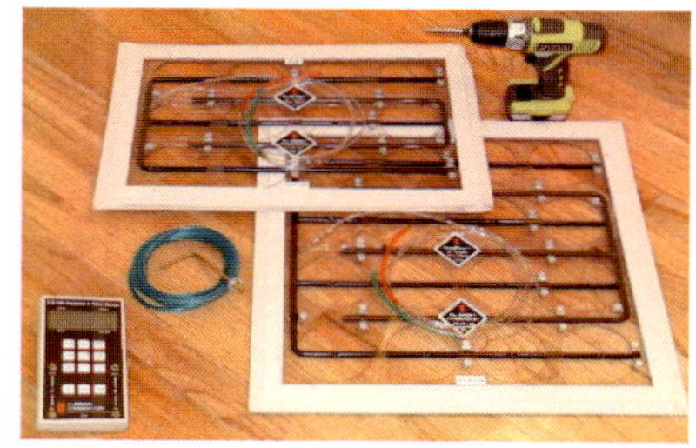

Materials Needed:

- flow plate (also called a flow grid)
- digital manometer
- short pressure hose
- static pressure probe (or Pitot tube)
- 12 or 18 volt drill and drill bit (7/32” to 1/4")

PREPARATION

Turn off the air handler according to A.02. Drill a 7/32” hole in the supply plenum, away from bends or turbulent features (or in the return plenum, if a good sampling location is not found).

Measure the filter slot or read the size on the filter (if it fits properly in the slot), and assemble the flow plate to fill the slot perfectly, according to flow plate manual. The extenders around the flow plate should be arranged on the inside of any sharp turns in the ductwork. Set the assembled flow plate aside. ***HEADS UP:** *This test will be much more accurate if the flow plate is installed at a central return register, rather than at a 180° turn in the airflow, as it is in my demonstration.*

If there is a central humidifier that incorporates a duct from supply plenum to return plenum, make sure the bypass damper is closed. You don't want the temperature of the return air changing during the test; that would also affect the air's density.

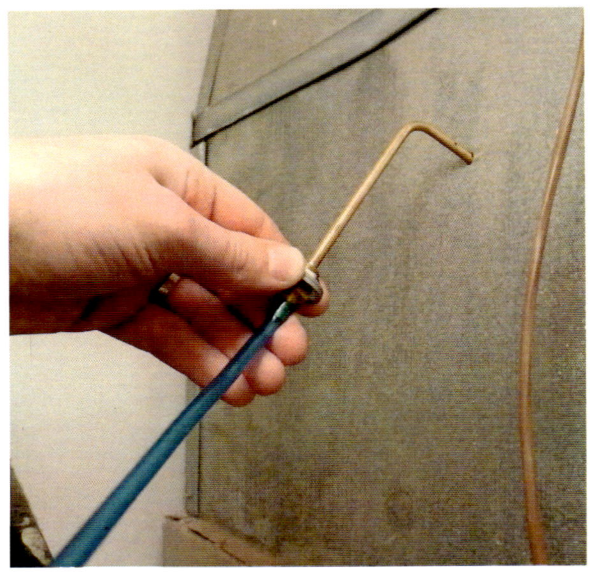

TESTING

Insert the static pressure probe fully into the drilled hole in the plenum, pointing into the airflow, and connect it to the Input tap of Channel A on the manometer with a pressure hose. If using a Pitot tube, connect the hose to the static pressure tap, not the total pressure tap (the static pressure tap is the larger diameter of the two taps).

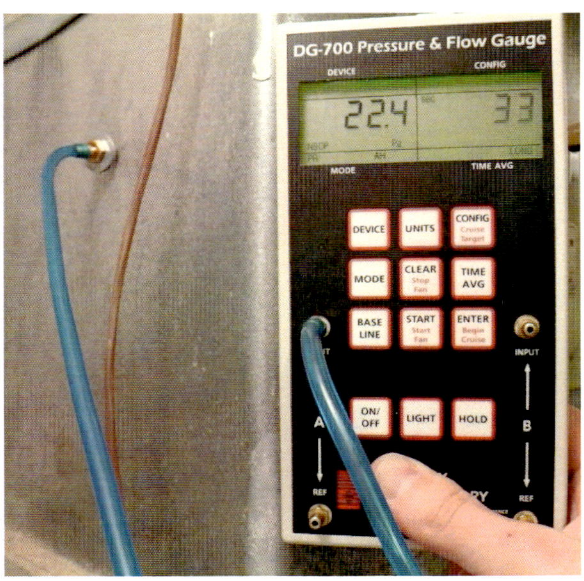

Turn on the air handler in cooling mode (*see A.02*). Set the manometer mode to PR/AH, and press 'START' to begin measuring a time average of the Normal System Operating Pressure (NSOP). To complete the baseline, press 'ENTER'. Or, use Pr/Pr mode and convert to airflow using tables in the flow plate's manual.

Turn off the air handler (*see A.02*), and carefully remove the filter.

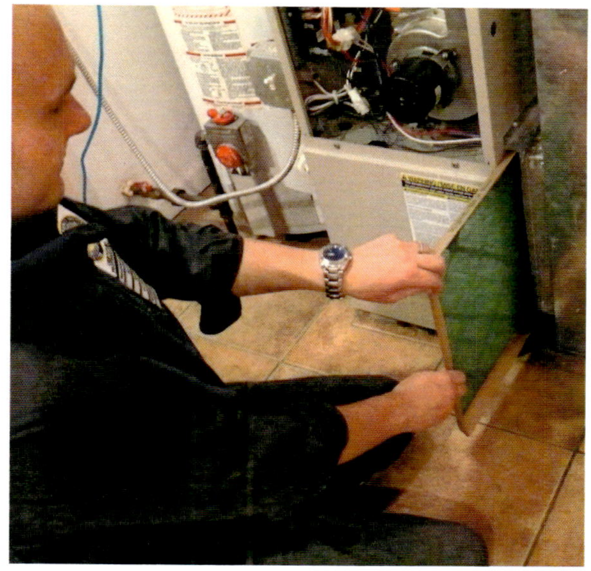

In its place, install the assembled flow plate. The extenders should go on the inside of any turns in the ductwork, so that the flow plate's sensors are able to measure the highest velocity airflows. ***TIP:** *Also try not to put extenders on the far side of the slot; it's a tight fit, and it might be a nightmare getting them out.*

Connect the flow plate's red pressure hose to Channel B input, and green hose to Channel B reference. Turn on the air handler again.

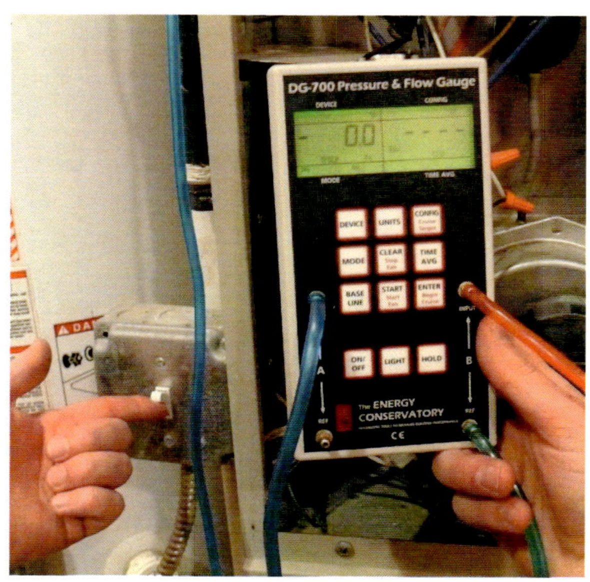

Wait for the air handler to reach full speed, and record the indicated airflow on Channel B. ***HEADS UP:** *If the air passing over the flow plate is not exactly 68°F, and you are not testing at sea level, use* **Figure 2.01.3** *to adjust for air density. Simply multiply the indicated airflow on the manometer by the Air Density Factor that corresponds with the actual temperature around the flow plate, and your elevation.*

Table C.1.c: **Air Density Factors to Convert from *Indicated* Flow to *Volumetric* Flow.**

Temp. of air through the Metering Plate (F)	Elevation (feet)										
	0	1000	2000	3000	4000	5000	6000	7000	8000	9000	10000
0	0.933	0.950	0.968	0.986	1.005	1.023	1.043	1.062	1.083	1.104	1.125
10	0.943	0.961	0.978	0.996	1.016	1.034	1.054	1.074	1.095	1.116	1.138
20	0.953	0.971	0.989	1.007	1.026	1.045	1.065	1.085	1.106	1.128	1.150
30	0.963	0.981	0.999	1.017	1.037	1.056	1.076	1.097	1.118	1.139	1.162
40	0.973	0.991	1.009	1.028	1.048	1.067	1.087	1.108	1.129	1.151	1.173
50	0.983	1.001	1.019	1.038	1.058	1.077	1.098	1.119	1.140	1.162	1.185
60	0.992	1.010	1.029	1.048	1.068	1.088	1.108	1.130	1.152	1.174	1.197
70	1.002	1.020	1.039	1.058	1.078	1.098	1.119	1.140	1.163	1.185	1.208
80	1.011	1.030	1.049	1.068	1.089	1.109	1.130	1.151	1.174	1.196	1.219
90	1.021	1.039	1.058	1.078	1.099	1.119	1.140	1.162	1.184	1.207	1.231
100	1.030	1.049	1.068	1.088	1.109	1.129	1.150	1.172	1.195	1.218	1.242
110	1.039	1.058	1.078	1.097	1.118	1.139	1.161	1.183	1.206	1.229	1.253
120	1.048	1.067	1.087	1.107	1.128	1.149	1.171	1.193	1.216	1.240	1.264
130	1.057	1.076	1.096	1.117	1.138	1.159	1.181	1.203	1.227	1.250	1.275
140	1.066	1.085	1.106	1.126	1.148	1.169	1.191	1.213	1.237	1.261	1.285
150	1.075	1.094	1.115	1.135	1.157	1.178	1.201	1.224	1.247	1.271	1.296

Volumetric Flow = Indicated Flow x Sqrt (0.075/air density) where air density is the density of air, in lbs/ft^3, going through the Metering Plate.

Figure 2.01.3 (excerpt printed with permission of Energy Conservatory)

2.01-C Vane Anemometer Method

Materials Needed:

- vane anemometer
- 12 or 18 volt drill and drill bit (3/4”)
- adhesive tape for open filter slot
- metal plugs or aluminum tape

PREPARATION

Turn off the air handler according to A.02. Prepare for a duct traverse by drilling three equidistant 3/4” holes in the straightest part of the return plenum, in a line perpendicular to the airflow, and away from bends or turbulent features. If the duct system is full of turbulent features, you may need to choose the 'least bad' place to drill the hole. Do your best.

Measure the cross-sectional area (width and height) of the return plenum at the testing location.

Configure the vane anemometer to calculate CFM through this cross-sectional area.

Ensure the filter slot is sealed, either with an installed filter slot cover or with tape. This ensures all the airflow will be measurable by your probe, and none will sneak in through the filter slot.

TESTING

Turn on the air handler in cooling mode (*see A.02*), and align the anemometer's probe so that it faces into the airflow (indicated by the calibration dot at the tip). Place your finger in alignment with this dot so you can keep the probe pointed in the right direction while it's inside the ductwork.

Extend the probe to the full depth of the plenum, and insert it completely in the first hole (keeping your finger in place for the proper facing). You'll feel the probe touch the far side of the duct. Put the anemometer into Timed Mean (or 'average') mode. If you prefer, use a Multi-Point Mean mode, using individual measurements in the center of equal squares throughout the cross section.

Begin taking a timed average reading (or a multi-point average), and pull the probe carefully out from the plenum as you do so. You'll find very high flow spots, and very low flow spots, which is why the average across the entire depth is critical.

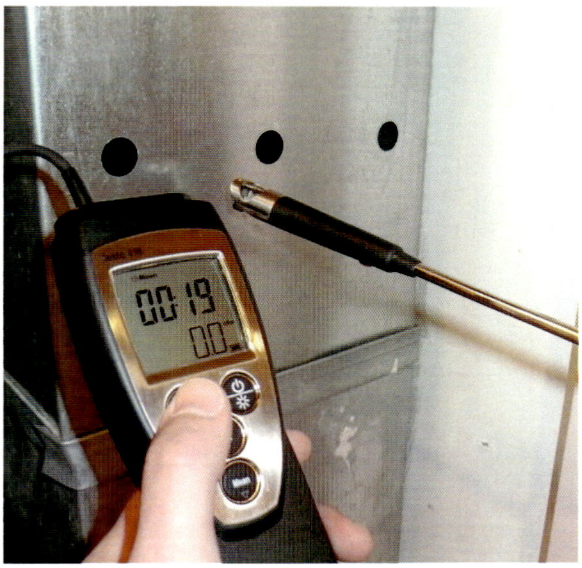

When the probe is about to emerge from the hole, pause your readings. Completely insert the probe straight into the next hole, until you feel it touch the far side.

Continue taking the Mean reading while pulling the probe out toward you, traversing the duct from one side to the other. Again, pause your readings when the probe is pulled to the edge of the hole, and insert the probe at the third hole to finish the duct traverse.

When you've finished the mean readings across the entire depth of all three testing holes, use the anemometer's Mean function to calculate the average airflow through the system.

Clean up by plugging your awesome holes with removable metal plugs or tape, for the next person who feels like traversing this plenum.

2.01-D Pressure Matching Method Using a Powered Flow Hood

Materials Needed:

- duct leakage-testing fan
- digital manometer
- short pressure hoses
- static pressure probe
- duct mask and/or plastic wrap
- cardboard at least as big as the air handler cabinet cover
- 12 or 18 volt drill and 7/32" drill bit

CONCEPT

The duct leakage-testing fan is already calibrated to measure precise airflows. Force the air handler to pull all of its air through the testing fan in order to measure system airflow. Of course, making your tiny testing fan the only opening available to the air handler means a restriction in the system airflow. That's why you always set up with the testing fan in pressurization mode; you'll need to run the testing fan to assist the air handler in getting all the air it can handle.

PREPARATION

Turn off the air handler according to A.02, and drill a 7/32" hole in the supply plenum, away from bends or turbulent features. Open a window or door to outdoors, to prevent pressure changes in the home during the test.

Determine where you'll install the duct leakage-testing fan. Get as close to the air handler as possible; mounting the fan directly to the air handler cabinet is possible by cutting a sheet of cardboard to fit. If the cabinet is not accessible, use a central return.

Seal any holes or penetrations into the blower cabinet, since the testing fan will be restricting the flow of the air handler. When the air handler gets starved for air, it will suck much harder through all the small holes and gaps available around it.

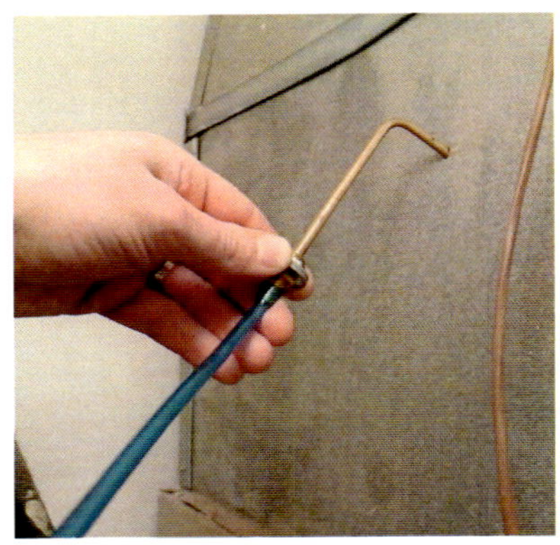

TESTING

Insert the static pressure probe into the drilled hole in the plenum, pointing into the airflow, and connect it to the input tap of Channel A on the manometer with a pressure hose.

Turn on the air handler in cooling mode (*see A.02*). Record the static pressure measured (in inches of water column or in Pascals). If the reading fluctuates significantly, use a longer time average (*see A.03*) or relocate the probe and measure again.

Turn off the air handler. Block airflow from the return plenum using the filter wrapped in plastic, or by inserting another air barrier.

Set up the duct leakage testing fan in pressurization mode directly on the air handler cabinet, or as close as you can get, so that *all* air pulled into the air handler goes through your testing fan first.

Run air handler again, and simultaneously run the duct-testing fan to assist it. Your goal is to reach the normal system operating static pressure and measure the flow through the duct blaster.

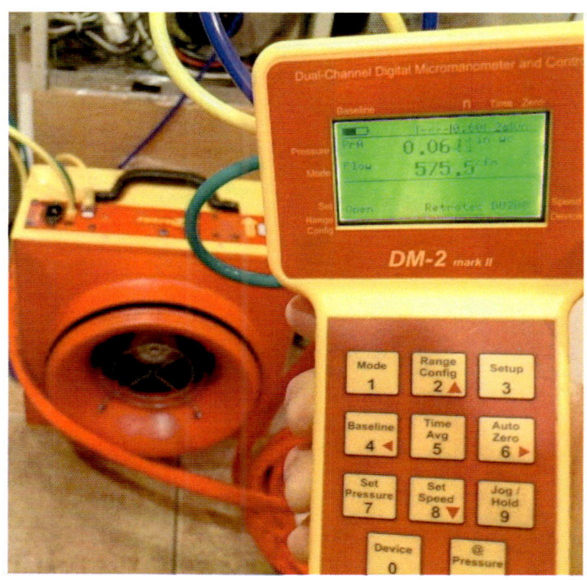

INTERPRETATION

If the Normal System Operating Pressure is not able to be duplicated in test configuration, even with the assistance of the testing fan, use **Equation 2.01.4** and the Tested Pressure Achieved (TPA) flow and pressure data to adjust for this difference**:**

Equation 2.01.4

$$\boldsymbol{CFM_{NSOP} = CFM_{TPA} \times \sqrt{\frac{NSOP}{TPA}}}$$

Where:

CFM_{NSOP} = Adjusted airflow to emulate airflow at NSOP [CFM]

CFM_{TPA} = Indicated airflow at the tested pressure achieved [CFM]

NSOP = Normal system operating pressure [inWC]

TPA = Actual pressure achieved in supply plenum [inWC]

Example:

CFM_{TPA} = 475.5 CFM

NSOP = .0923inWC

TPA = 0.0641inWC

$$\boldsymbol{CFM_{NSOP} = 575.5 \times \sqrt{\frac{0.0923}{0.0641}} = 691\ CFM\ at\ NSOP}$$

2.01-E Temperature Rise Method

Materials Needed:

- two probe thermometers (or single dual-input thermometer)
- 12 or 18 volt drill and drill bit

CONCEPT

Measuring temperature change before and after the heat exchanger of a furnace is another way of diagnosing system airflow, though this is a less accurate method than the others. The temperature rise (ΔT) should be right in the midst of the range the manufacturer specifies; if it's greater, the airflow may be too low. If the ΔT is much smaller than expected, then the airflow may be too high. For this test, you'll need to fire the furnace *(see A.02 for further reading on controlling the HVAC equipment).* The more readings you can take inside the ductwork, the better for accuracy; there are warm and cool spots, and you'll want to average the temperatures you measure as you traverse the duct.

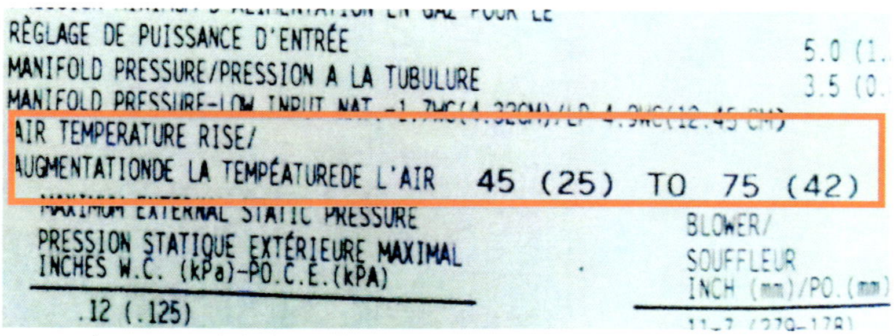

PREPARATION

Drill a hole for the thermometer (or multiple holes, if you're using a short thermometer probe) in the supply plenum, away from the A/C coil and any other obstructions to the flow inside the ductwork. ***TIP:** *Put the testing holes out of line-of-sight with the heat exchanger, to avoid measuring any radiant heat from it.*

If there is a central humidifier that incorporates a duct from supply plenum to return plenum, make sure the bypass damper is closed. You don't want the temperature of the return air changing during the test; that would also affect the air's density.

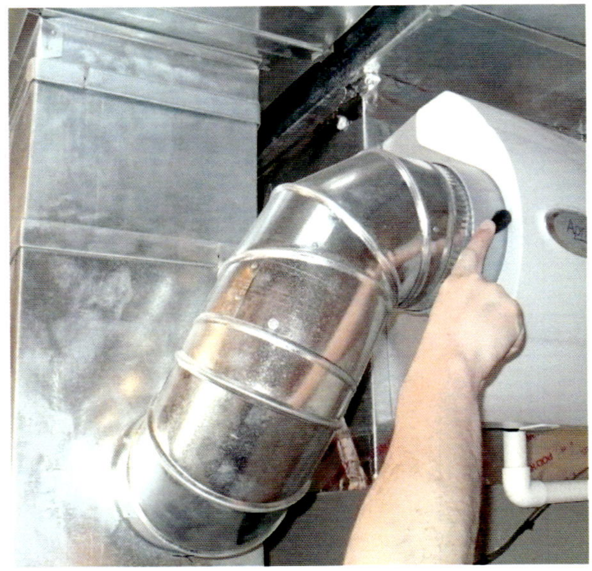

TESTING

Fire the furnace (*see A.02*), and let it reach steady state efficiency (when the temperature levels out, usually 5-10 minutes). Measure the temperature at multiple points in the return and supply plenum simultaneously (staying out of line-of-sight with the heat exchanger). If you measure them one at a time, the temperatures may change, and your ΔT measurement will be inaccurate.

The temperature difference (ΔT) between the return and supply plenums is the temperature rise. A dual-input thermometer will calculate the ΔT directly. It's important to take multiple readings.

INTERPRETATION

Average the multi-point readings. Compare your test data to the manufacturer's specification for temperature rise on the equipment label. To be accurate in your calculation, test the actual furnace input by clocking the gas or electric meter using 2.14 or 3.05. Use **Equation 2.01.5** to calculate the system airflow through a furnace based on temperature rise:

Equation 2.01.5

$$CFM_{Duct\ System} = \frac{Furnace\ Output}{1.08 \times \Delta T}$$

Where:

$CFM_{Duct\ System}$ = Total system airflow [CFM]

Furnace Output = Btu/h Input X Efficiency [Btu/h] (*verify with 2.14 or 3.05*)

ΔT = Measured Temperature Rise [°F]

Example:

Furnace Output = 58,650 Btu/h x .8 Efficiency = 46,920 Btu/h

ΔT = 125.3°F supply temp, 68.9°F return temp = 56.4°F Temperature Rise

$$CFM_{Duct\ System} = \frac{46920}{1.08 \times 56.4} = 770\ CFM$$

QUIZ System Airflow

1) When measuring airflow with a Pitot tube, why is it important to seal off the filter slot? If this step were not taken, what would happen to your resulting airflow measurement? ____

____

____

2) Why is it best to measure system airflow in cooling mode? ______________

____

3) What is density of the air in a return plenum given that it is located in Seattle Washington (Elevation of 450 ft), and the return plenum temperature is 68° F? ____

____

____

4) Using the Pitot tube, you traversed a return plenum twice with velocity pressures of P_{V1}= 0.082 IWC, and P_{v2}=0.087. You have established that the density of the air is 0.0786 lbs/ft^3. What is the velocity of the air in the return plenum? ____

____

____

5) You have measured the cross sectional area of the ductwork and found it to be 10" x 24". Assuming a velocity of 1000 ft/min, what volume of air is moving through the system? If there is a 3.5-ton condenser that is part of this system, is the airflow sufficient? ______________________________

____

____

6) You measured the airflow through an air handler using a Flow Plate and recorded 834 CFM. You noted that the temperature of the air was 70° F and you are located in Blacksburg, VA (Elevation 2500 ft). What is the adjusted airflow through the system? ____________________

____

____

7) When performing system airflow testing using the Pressure Matching Method with a Powered Flow Hood, you measure a normal system operating pressure of 0.0876 IWC in the supply plenum. After setting up the powered flow hood, you are getting an airflow reading of 639 CFM at a test pressure of 0.0799 IWC. What is the adjusted airflow to reflect the airflow at normal system operating pressure? ____________________

____

____

____

8) A 75 kBtu furnace that is 92.5% efficient has a supply temperature of 132.9 ° F and a return temperature of 67.4° F. What is the estimated airflow over the heat exchanger? ____________________

____

____

____

Discussion and Critical Thinking:

1) Discuss the behavior of laminar and turbulent flow. How would this relate to system airflow testing? What ductwork conditions would create turbulent flow? How about laminar flow? ____________________

____

____

2) What are some reasons that you would want to close the humidifier damper between a return and supply plenum when conducting system airflow testing? ____

____

____

3) When traversing a duct, what is the fundamental difference between multipoint testing and time averaged testing? ______________________

____

____

4) Discuss the pros and cons to each method of system airflow measurement. ____

____

____

____

____

2.02 Total External Static Pressure

Materials Needed:

- static pressure probe or Pitot tube
- digital manometer
- 12 or 18 volt drill and drill bit

CONCEPT

Static pressure is resistance to airflow, created by the size, layout, and texture of the ductwork. Just like the human circulatory system, the ductwork's flow and pressure are closely related: higher blood pressure results in less bloodflow, just as higher static pressure results in less airflow. The External Static Pressure is the pressure of everything external to the air handler, and must be measured as close before and after the air handler as possible. The more obstructions to airflow there are upstream and downstream, the harder the fan will be pulling and pushing on the air just behind and ahead of it. Components installed inside the ductwork (heat exchanger, A/C coil, filter) increase the static pressure close to the fan, as do undersized or rough ductwork, or a kinked or overly twisted layout.

Measuring static pressure when the A/C coil has been in use means the reading will be higher, since condensation on the coil creates resistance to airflow. Therefore, this test should be performed when the coil is dry.

Each side of the air handler fan will have a static pressure; the supply side pressure is positive, and the return side pressure is negative. When measuring total static pressure, including both the supply and return, *count both sides as being positive* and add them together to get a bigger number than either of the individual sides.

Static pressures are normally measured in inches of water column (inWC), and should typically be less than 0.25inWC on either the supply or return side, with a total pressure of 0.5inWC or less. It is possible to verify total system airflow using total external static pressure, as well, if you have a fan curve from the HVAC equipment manufacturer. The fan curve table will show the direct relationship between static pressure in inWC and airflow in CFM.

PREPARATION

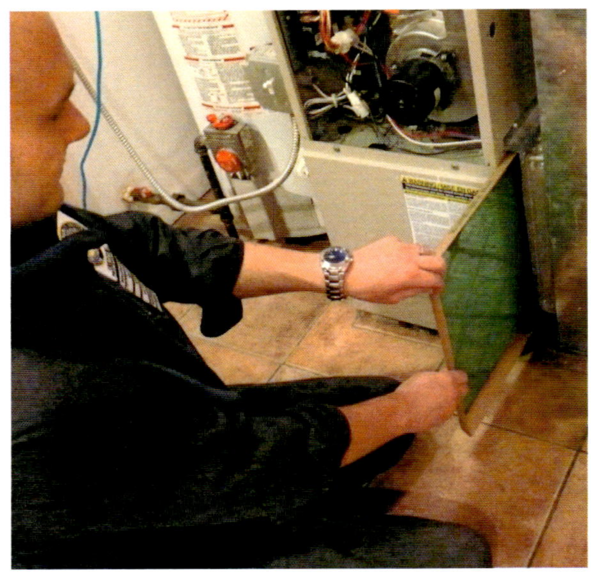

Inspect the filter and A/C evaporator coil (if accessible) and make sure both are clean and dry. Any buildup of dust or moisture will result in a higher static pressure reading.

Locate or drill a hole for the static pressure probe on the return side of the air handler, but still between the fan and the filter. The only appropriate testing location may therefore be in the blower cabinet itself. Usually you can find a wiring penetration into the blower cabinet that will work for your test, in which case no drilling is necessary.

Locate or drill a hole for the static pressure probe past the heat exchanger of the furnace, but before the A/C evaporator coil. In many gas furnace assemblies, there is no space in the supply plenum between the end of the furnace and the beginning of the evaporator coil. The only appropriate testing location may therefore be in the furnace's heat exchanger cabinet.

TIP:** *If you're hard pressed for a testing location before the evaporator coil, and you're feeling adventurous, you can try getting into the airflow around the primary heat exchanger around the high temperature limit switch.* ***WARNING: *You may be about to short out the electrical system accidentally. If in any doubt about your ability, choose another 'least bad' location.*

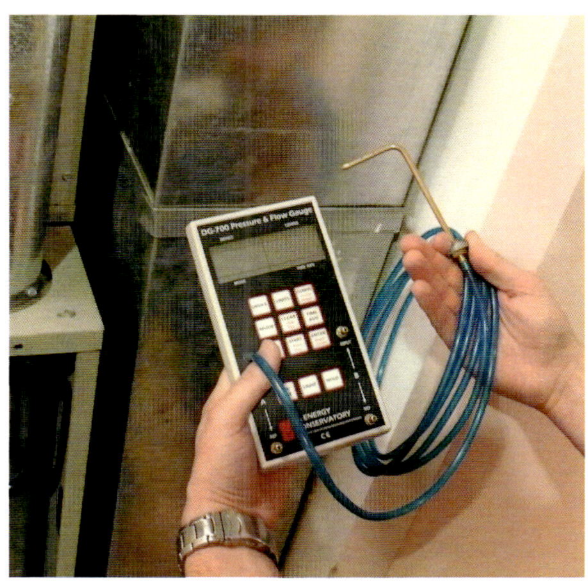

Connect the static pressure probe to one end of the pressure hose, and connect the other end to either manometer channel's Input tap. Set up the manometer to measure pressure (inWC) in the ductwork with reference to the home (*see A.03*).

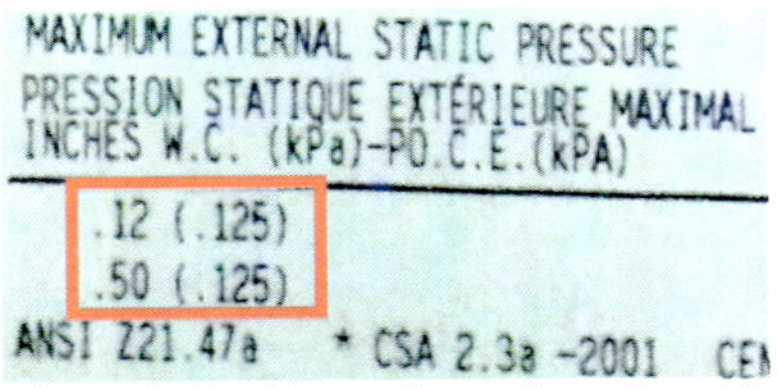

Refer to the manufacturer's documentation for the air handler's designed maximum static pressure. This may be difficult to find, and may require contacting the manufacturer; however, you don't need the documentation before testing, you can refer to it later.

TESTING

Insert the static pressure probe fully into the hole on the return side of the air handler, pointing into the airflow. **HEADS UP:** *If your testing location is inside the blower cabinet, you may encounter bizarrely high and low pressures. Fish around for an average reading.*

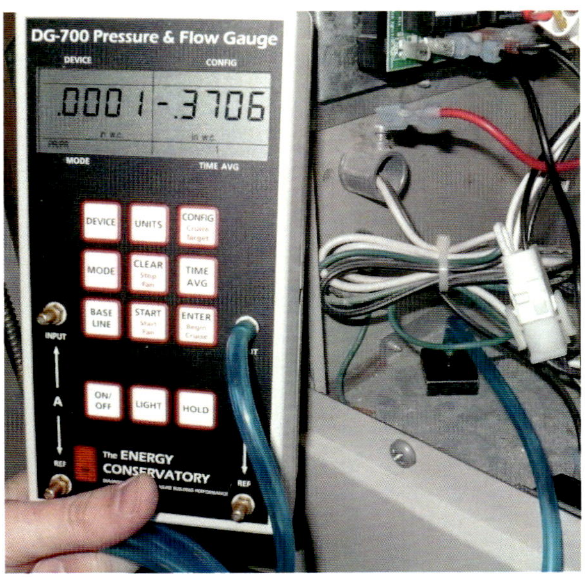

Turn on the air handler (*see A.02*) and record the return static pressure in inWC shown on the manometer. Though it shows a negative, record it as a positive number.

Remove the static pressure probe from the return side, and insert it at the hole on the supply side of the air handler, pointing into the airflow. Record the static pressure of the supply side.

INTERPRETATION

Add the supply and return static pressures together to calculate the total external static pressure. If the total is higher than the manufacturer's designed maximum, as it is in our example, the duct system is too restrictive, and airflow is being impeded. If this is the case and you are not an HVAC professional, hand off the information to an expert who can investigate further and recommend a solution.

QUIZ Total External Static Pressure

1) Define what Total External Static Pressure is and where it should ideally be measured. ______________________________

____

____

2) You find that the external static pressure is 0.267 IWC on the supply side and -0.310 on the return side. The manufacturer specifies that Total External Static Pressure should not exceed 0.650 IWC. Is this system in compliance? ______________________________

____

____

____

Discussion and Critical Thinking:

1) How does a dirty filter, dirty A/C coil, or a wet A/C coil affect total external static pressure? ______________________________

____

____

____

2) Discuss various dangers of drilling into the furnace cabinet, or removing installed parts for a testing location. ______________________________

____

____

____

2.03 Temperature Rise/Drop

CONCEPT

The amount of heat that is added to air by a furnace (or removed from it by an air conditioner) is an important measure of performance. The furnace's temperature difference (ΔT) before and after the heat exchanger is called temperature rise. Temperature drop, or 'temperature split', is the ΔT over an air conditioner's evaporator coil.

Temperature rise of a furnace is relatively straightforward, through comparison of the dry bulb temperatures (or 'sensible heat') of the supply and return sides.

Temperature drop is more complicated, since it lowers both the sensible heat and the latent heat, or humidity. The total heat of both combined is called 'Enthalpy'.

2.03-A Temperature Rise Method

2.03-B Temperature Drop Method

2.03-A Temperature Rise Method

Materials Needed:

- two probe thermometers or single dual-input thermometer
- 12 or 18 volt drill and drill bit

If the temperature change (ΔT) is much larger than the manufacturer's design, then the airflow may be too low. If the ΔT is much smaller than expected, then the airflow may be too high. You should expect a reading right in the middle of the range given by the manufacturer on the label, which in this case will be 60°F.

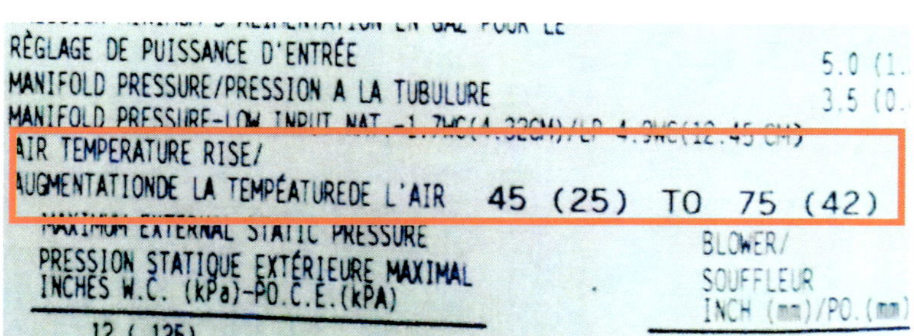

For this test, you'll need to fire the furnace (*see A.02 for further reading on controlling the HVAC equipment*).

The more readings you can take inside the ductwork, the better for accuracy; there are warm and cool spots, and you'll want to average the temperatures you measure as you traverse the duct.

PREPARATION

Drill a hole for the thermometer (or multiple holes, if you're using a short thermometer probe) in the return plenum and in the supply plenum, away from the A/C coil and any other obstructions to the flow inside the ductwork. ***TIP:** *Try to put the testing holes out of line-of-sight with the heat exchanger, since you don't want to measure any of the radiant heat flow from it. You may have to choose a 'least bad' location, but if you're using a probe thermometer like ours, the actual temperature sensor is hidden out of line-of-sight by the probe housing, which helps.*

TESTING

Fire the furnace (*see A.02*), and let it reach steady state efficiency (when the temperature levels out, usually 5-10 minutes). Measure the temperature in the return simultaneously with a series of points supply plenum. If you measure them one at a time, the temperatures may change, and your ΔT measurement will be inaccurate.

The temperature difference (ΔT) between the return and supply plenums is the temperature rise. A dual-input thermometer will calculate the ΔT directly, which you'll use to average the multiple-point readings.

INTERPRETATION

Average the multi-point readings. Compare your test data to the manufacturer's specification for temperature rise on the equipment label.

2.03-B Temperature Drop Method

Materials Needed:

- two probe wet bulb thermometers or single dual-input wet bulb thermometer
- 12 or 18 volt drill and drill bit

The design temperature split over an air conditioning evaporator coil is generally ~21°F, with 75% of the capacity going into lowering the sensible heat (or temperature), and 25% going into lowering the latent heat (or relative humidity). Since the A/C lowers both temperature and humidity, it's important to measure both when measuring temperature split, in order to diagnose actual capacity. **Figure 2.03.1** shows ideal temperature split:

Figure 2.03.1

Return Air Dry-Bulb (°F) ($T_{return, db}$) \ Return Air Wet-Bulb (°F) ($T_{return, wb}$)	50	51	52	53	54	55	56	57	58	59	60	61	62	63	64	65	66	67	68	69	70	71	72	73	74	75	76
70	20.9	20.7	20.6	20.4	20.1	19.9	19.5	19.1	18.7	18.2	17.7	17.2	16.5	15.9	15.2	14.4	13.7	12.8	11.9	11.0	10.0	9.0	7.9	6.8	5.7	4.5	3.2
71	21.4	21.3	21.1	20.9	20.7	20.4	20.1	19.7	19.3	18.8	18.3	17.7	17.1	16.4	15.7	15.0	14.2	13.4	12.5	11.5	10.6	9.5	8.5	7.4	6.2	5.0	3.8
72	21.9	21.8	21.7	21.5	21.2	20.9	20.6	20.2	19.8	19.3	18.8	18.2	17.6	17.0	16.3	15.5	14.7	13.9	13.0	12.1	11.1	10.1	9.0	7.9	6.8	5.6	4.3
73	22.5	22.4	22.2	22.0	21.8	21.5	21.2	20.8	20.3	19.9	19.4	18.8	18.2	17.5	16.8	16.1	15.3	14.4	13.6	12.6	11.7	10.6	9.6	8.5	7.3	6.1	4.8
74	23.0	22.9	22.8	22.6	22.3	22.0	21.7	21.3	20.9	20.4	19.9	19.3	18.7	18.1	17.4	16.6	15.8	15.0	14.1	13.2	12.2	11.2	10.1	9.0	7.8	6.6	5.4
75	23.6	23.5	23.3	23.1	22.9	22.6	22.2	21.9	21.4	21.0	20.4	19.9	19.3	18.6	17.9	17.2	16.4	15.5	14.7	13.7	12.7	11.7	10.7	9.5	8.4	7.2	5.9
76	24.1	24.0	23.9	23.7	23.4	23.1	22.8	22.4	22.0	21.5	21.0	20.4	19.8	19.2	18.5	17.7	16.9	16.1	15.2	14.3	13.3	12.3	11.2	10.1	8.9	7.7	6.5
77	-	24.6	24.4	24.2	24.0	23.7	23.3	22.9	22.5	22.0	21.5	21.0	20.4	19.7	19.0	18.3	17.5	16.6	15.7	14.8	13.8	12.8	11.7	10.6	9.5	8.3	7.0
78	-	-	-	24.7	24.5	24.2	23.9	23.5	23.1	22.6	22.1	21.5	20.9	20.2	19.5	18.8	18.0	17.2	16.3	15.4	14.4	13.4	12.3	11.2	10.0	8.8	7.6
79	-	-	-	-	-	24.8	24.4	24.0	23.6	23.1	22.6	22.1	21.4	20.8	20.1	19.3	18.5	17.7	16.8	15.9	14.9	13.9	12.8	11.7	10.6	9.4	8.1
80	-	-	-	-	-	-	25.0	24.6	24.2	23.7	23.2	22.6	22.0	21.3	20.6	19.9	19.1	18.3	17.4	16.4	15.5	14.4	13.4	12.3	11.1	9.9	8.7
81	-	-	-	-	-	-	-	25.1	24.7	24.2	23.7	23.1	22.5	21.9	21.2	20.4	19.6	18.8	17.9	17.0	16.0	15.0	13.9	12.8	11.7	10.4	9.2
82	-	-	-	-	-	-	-	-	25.2	24.8	24.2	23.7	23.1	22.4	21.7	21.0	20.2	19.3	18.5	17.5	16.6	15.5	14.5	13.4	12.2	11.0	9.7
83	-	-	-	-	-	-	-	-	-	25.3	24.8	24.2	23.6	23.0	22.3	21.5	20.7	19.9	19.0	18.1	17.1	16.1	15.0	13.9	12.7	11.5	10.3
84	-	-	-	-	-	-	-	-	-	25.9	25.3	24.8	24.2	23.5	22.8	22.1	21.3	20.4	19.5	18.6	17.6	16.6	15.6	14.4	13.3	12.1	10.8

Enthalpy is the total measurable change in both sensible and latent heat that you're ultimately after in a temperature drop diagnostic. The change in enthalpy is designed always to be exactly 6.667 Btus per pound of air at design airflow. Be sure to measure the actual total system airflow (*see 2.01*) before attempting this test. You'll need to consult **Figure 2.03.2** to convert wet-bulb temperature to enthalpy, or you can use the psychrometric calculator at:

www.BuildingPerformanceWorkshop.com/calculator

Figure 2.03.2

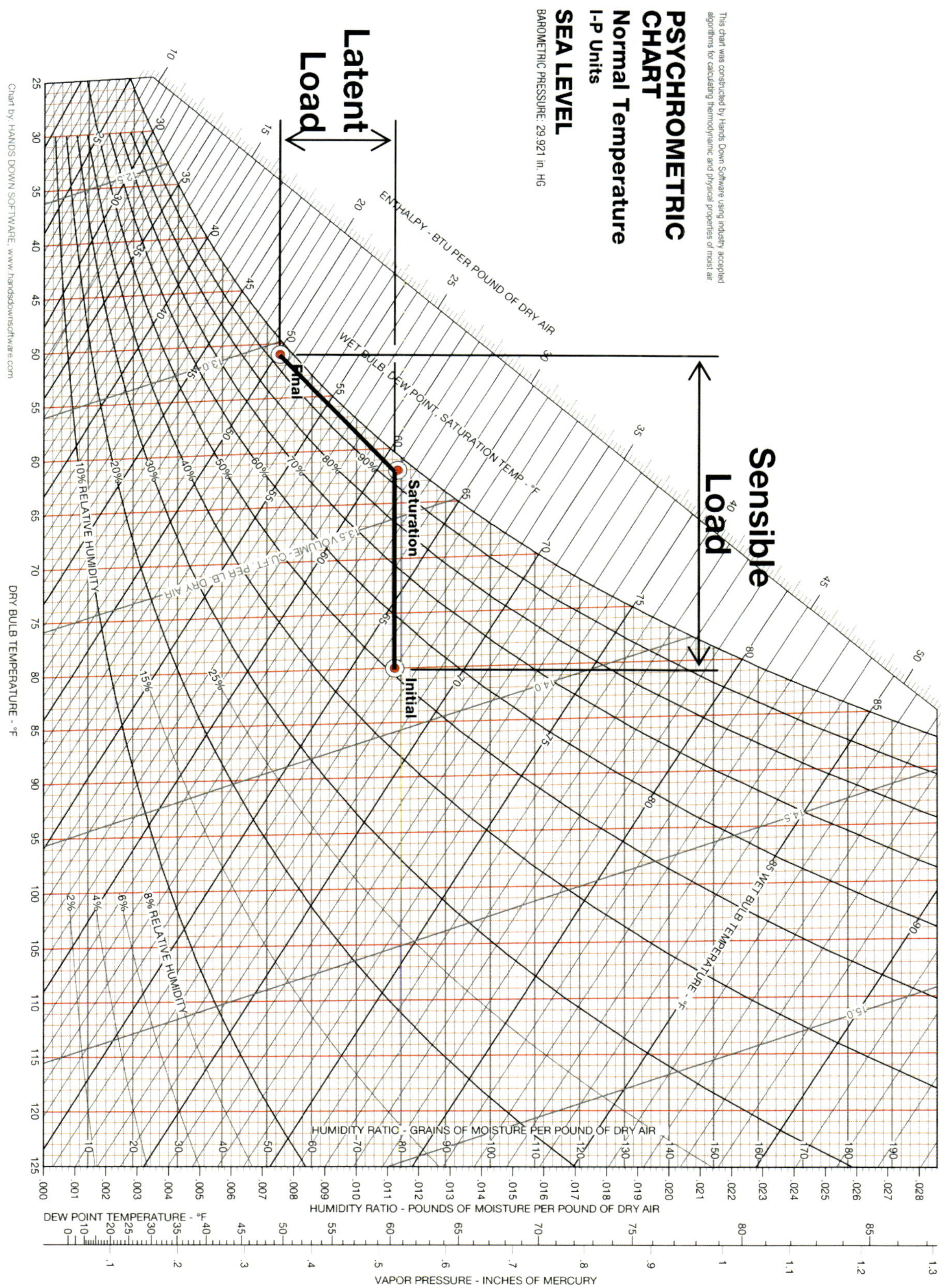
This chart was constructed by Hands Down Software using industry accepted algorithms for calculating thermodynamic and physical properties of moist air
PSYCHROMETRIC CHART
Normal Temperature
I-P Units
SEA LEVEL
BAROMETRIC PRESSURE: 29.921 in. HG
Latent Load
Sensible Load
Final
Saturation
Initial
ENTHALPY - BTU PER POUND OF DRY AIR
WET BULB, DEW POINT, SATURATION TEMP - °F
13.5 VOLUME - CU.FT. PER LB. DRY AIR
10% RELATIVE HUMIDITY
8% RELATIVE HUMIDITY
85 WET BULB TEMPERATURE - °F
HUMIDITY RATIO - GRAINS OF MOISTURE PER POUND OF DRY AIR
HUMIDITY RATIO - POUNDS OF MOISTURE PER POUND OF DRY AIR
DRY BULB TEMPERATURE - °F
DEW POINT TEMPERATURE - °F
VAPOR PRESSURE - INCHES OF MERCURY
Chart by: HANDS DOWN SOFTWARE, www.handsdownsoftware.com

PREPARATION

Drill a hole for the thermometer (or multiple holes, if you're using a short thermometer probe) in the supply plenum, away from the A/C coil and any other obstructions to the flow inside the ductwork. ***HEADS UP:** *Try to put the testing holes out of line-of-sight with the A/C coil, since you don't want to measure any of the radiant heat flow into it. You may have to choose a 'least bad' location, but if you're using a probe thermometer like ours, the actual temperature sensor is hidden out of line-of-sight by the probe housing, which helps.*

TESTING

Turn on the A/C (*see A.02*), and let it run for 10 minutes to reach steady state efficiency. Then measure the wetbulb temperature in the return and supply simultaneously. Readings in multiple locations increase the test's accuracy. Average the multi-point wet bulb readings.

INTERPRETATION

The A/C system's design capacity is 12,000 Btu per ton (e.g., a 3 ton unit should move 36,000 Btu per hour). The system may be performing below or beyond capacity in actuality. Since we know the actual total system airflow is correct, we're ready to diagnose the system's performance based on the temperature split.

Use **Equation 2.03.1** to calculate the actual Btu per hour of the A/C system:

Equation 2.03.1 $$\boldsymbol{Btu/h = 4.5 \times CFM \times \Delta h}$$

Where:

CFM = Tested total system airflow [CFM]

Δh = Change in Enthalpy [Btu/lb of air]

Example: CFM = 808 CFM tested on a 2 ton A/C (should be moving 24,000 Btu/hr) Δh = Return wet bulb 69.6°F, supply wet bulb 62°F = 5.84 Btu/lb of air $\boldsymbol{Btu/h = 4.5 \times 808 \times 5.84 = 21,234\ Btu/h}$

QUIZ Temperature Rise/Drop

1) When considering temperature rise, why is it safer to measure supply temperatures in various locations? If taking multiple measurements, which measurement do you choose to use for your calculation? ________________

____

____

2) The measured supply plenum temperature is 139.6° F and the measured return plenum temperature is 62.9° F. What is the temperature rise of this furnace? If the furnace is specified to have a temperature rise of 45-75° F, is the temperature rise acceptable? ________________

____

____

3) After measuring the temperature rise of the furnace, you find that the measured value is a few degrees less than what the furnace specifies. What could be a likely cause of this? ________________

____

____

4) What is the ideal temperature split of a system that has a Return Air wet bulb of 69° F, and a Return Air Dry bulb reading of 74° F? ________________

____

____

5) Using a psychometric calculator, Determine the change in enthalpy (Δh) given the following conditions: Altitude = 685 Ft, Supply wet bulb = 52.5° F, Return wet bulb = 65.9 °F, 50% RH. Is this higher than lower than what you would expect? ________________

____

6) Using **Figure 2.03.2**, what is the change of enthalpy between a 75° F wet bulb reading and a 65° F wet bulb reading? ______________________

____

7) A 3-ton system has the following parameters: 1167 CFM airflow, supply wet bulb of 58.5 °F, and a return wet bulb of 68.0° F. Given the size of the system, is it performing at optimal capacity? Assume the test is performed at sea level. ______________________

____

8) Why is it important to know what the actual (measured) airflow is prior to performing this test? What are we trying to diagnose by performing a temperature drop test? ______________________

____

Discussion and Critical Thinking:

1) Discuss the differences between sensible and latent heat and how they relate to temperature rise and temperature drop. ______________________

____

2) How does airflow affect temperature rise? Discuss how temperature rise can be used as an indicator of airflow through a system. ______________________

____

3) What is the key difference between wet bulb and dry bulb temperatures?

____

____

2.04 Room by Room Pressure/Flow Balancing

CONCEPT

Balancing of the forced air system is critical for controlling comfort and air quality. That's why it's important to test the flows through individual supply and return registers, and their pressure effects on rooms. Register flows will be demonstrated here, and *for room pressures, refer to Method 1.03-B HVAC Pressure Effects for Zonal Pressure Testing.*

The measured room flows should be compared with a room-by-room airflow calculation (Manual J or equivalent).

2.04-A Vane Anemometer Method

2.04-B Flow Hood Method

2.04-C Plastic Bag Method

2.04-D Powered Flow Hood Method

2.04-A Vane Anemometer Method

Materials Needed:

- vane anemometer (larger vane is more accurate)

PREPARATION

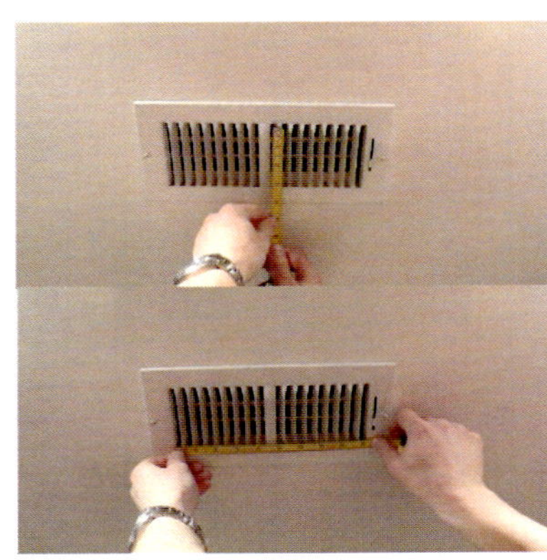

Measure the net free area of the register (the actual size of the openings, not the outer edge of the register). Remove any obstructions in front of the register to be tested, and open the damper completely. Remember where the damper was originally set; you'll want to return it to its original position once the register has been tested.

Calculate the square inches of the net free area and configure the vane anemometer to measure CFM for that size opening. If testing a return register, no K-Factor is necessary; if testing a supply register, the K-Factor will adjust for the resistance of the grill, and should be used in the configuration of the vane anemometer.

TESTING

Turn on the air handler (*see A.02*) and hold the vane anemometer close to the register to start it spinning. Set the anemometer to timed average or multi-point average.

Begin taking measurements of the register's airflow, steadily traversing the entire net free area. Stop the timed or multi-point testing and calculate the average airflow through the register.

2.04-B Flow Hood Method

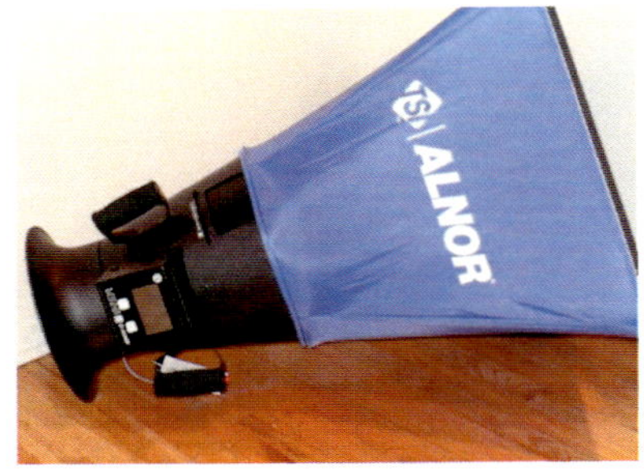

Materials Needed:

- flow hood (also called a balometer)

PREPARATION

Remove any obstructions in front of the register to be tested, and open the damper completely. Remember where the damper was originally set; you'll want to return it to its original position once the register has been tested.

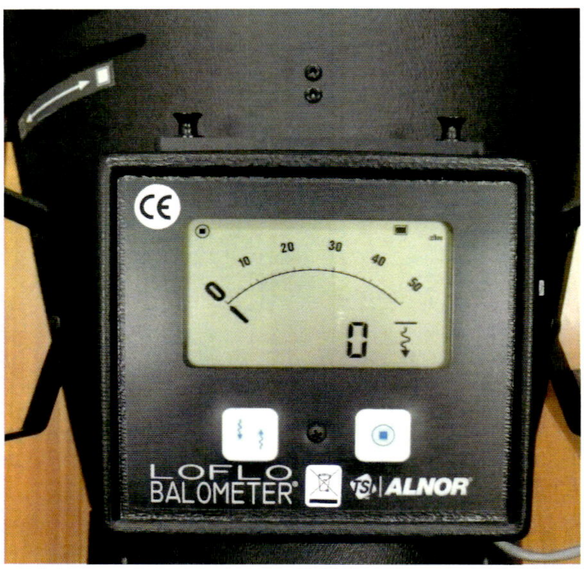

Configure the balometer for vent-closed operation, and for the direction of the airflow being tested. For airflows above 150 CFM, use vent open mode to adjust for the resistance of the flow hood.

TESTING

Place the flow hood firmly around the register, and record the airflow in CFM.

2.04-C Plastic Bag Method

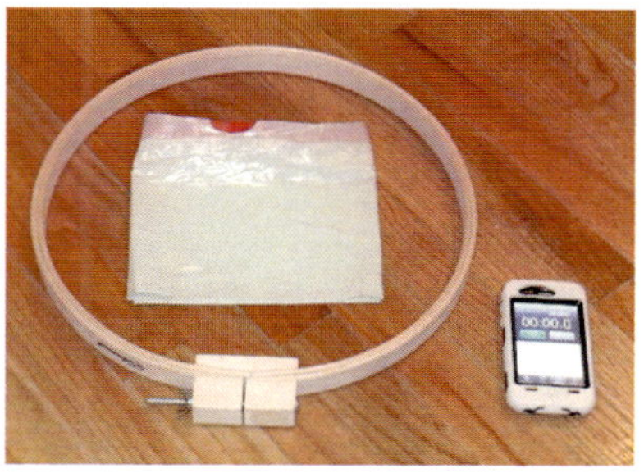

Materials Needed:

- plastic garbage bag
- frame for bag (needlepoint frame works well)
- stopwatch

PREPARATION

Clear the area around the register to be tested, and open the damper. Affix the bag to the frame to create an airtight seal. Be sure to have chosen a plastic bag that is both large and relatively lightweight; the longer it takes to fill, with the least resistance, the more accurate your measurement will be.

TESTING

Squeeze the air completely out of the bag, and start the stopwatch at the same time you cover the register with the bag.

Stop the clock as soon as the bag becomes completely filled with air. Record the number of seconds it took to fill the bag, and repeat the process for a minimum of 3 measurements. Average the measurements, and use **Equation 2.04.1** to calculate the airflow through the register.

INTERPRETATION

To convert the bag size to cubic feet, use the following conversion:

$$1 \text{ Gal [US]} = 0.134 \text{ ft}^3$$

Equation 2.04.1 $$\boldsymbol{Airflow_{Register} = \frac{Vol_{Bag} \times 60}{T}}$$

Where:

$Airflow_{Register}$ = Volumetric flow rate supplied to the register [CFM]

Vol_{Bag} = Volume of bag [ft^3]

T = Time elapsed for bag to fill [seconds]

Example:

Our 30-gallon bag was filled in 2.8 seconds, 3.2 seconds, and 2.7 seconds.

T = the average of the 3 readings is 2.9 seconds.

Vol_{Bag} = 30 gallons x .134 ft^3 = 4 ft^3

$$\boldsymbol{Airflow_{Register} = \frac{4 \times 60}{2.9} = 83\ CFM}$$

2.04-D Powered Flow Hood Method

Materials Needed:

- duct leakage testing fan with flex duct snorkel
- digital manometer
- short pressure hoses
- large five-sided box (cardboard works fine)

PREPARATION

Remove any obstructions in front of the register to be tested, and open the damper completely. Remember where the damper was originally set; you'll want to return it to its original position once the register has been tested.

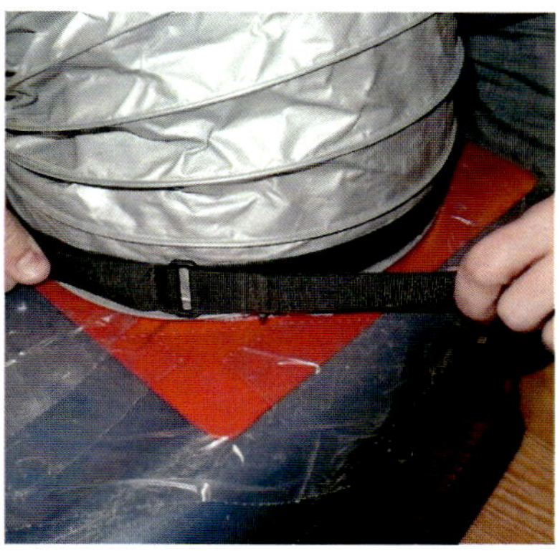

Cut a hole the same diameter as the snorkel's flex duct in the box, across from the open end. Tape the square flange to the box's opening, and attach to it one end of the snorkel. Make the box airtight except for the wide-open sixth side, which should be able to fit tightly against the flat surface around the register to be tested.

Attach the other end of the snorkel to the duct leakage-testing fan. The testing fan should be arranged to move air in the same direction as the register to be tested (for example, if testing supply registers, set the fan up in depressurization mode, and if testing return registers, set up in pressurization mode). Install the flow ring on the fan that works best for the expected airflow (for example, if measuring a supply register, a low flow ring is needed to measure the 50-100 CFM flow, and if measuring a central return, leave the fan wide open).

Connect the duct-testing fan to the manometer's Channel B (post-fan tap to Input, and pre-fan tap to Reference) (*see A.03, 'Manometer Mechanics'*) and put the manometer in Pressure/Flow mode.

Drill an access hole (or install a pressure tap as I've done) for the pressure hose to measure the pressure inside the box during testing. Connect one end of the pressure hose to the inside of the box, and the other end to the Input tap on Channel A of the manometer.

TESTING

Place the box's open end firmly around the register, and begin measuring the pressure in the box with reference to the home on Channel A of the manometer.

Start the testing fan, and increase the speed until the pressure in the box with reference to the home is 0 Pa. At this point, the airflow entering the box is equal to the airflow leaving the box, and you have confirmed the flow through the register. Record the airflow on Channel B.

AUTHOR'S NOTE: I don't know if you can tell by my face, but this test is sort of a pain in the @#$. I recommend avoiding it, due to overall awkwardness (not what we call in our training center 'testing sexy'). If you already own a $2500 duct testing fan and your client needs one or two outlet flows tested (say, on an ERV), go for it. But I would hesitate to even attempt moving around a home with all of this equipment. To make it easier, the Energy Conservatory has developed a conversion kit for their Duct Blaster® that incorporates the box and a battery pack, which makes it more mobile, though this is brand new as of the writing of this edition.

QUIZ Room-by-Room Pressure/Flow Balancing

1) What is the purpose of a k-factor of a supply register when measuring flow with a vane anemometer? ____

____

2) A supply registers has the following properties: Overall dimensions – 6" x 12", Area of opening – 4" x 10", K-factor of 0.80. What is the area that would be entered into a vane anemometer prior to testing? ____________

____

____

3) When performing flow balancing with a vane anemometer, would multipoint readings or time averaging generally be more accurate? Why? ____________

____

____

4) Although relatively simple, list a few reasons why the 'plastic bag method' is prone to error. ____

____

____

5) After measuring the flow of a register with a balometer and vane anemometer, you conclude that the flow rate is 122 CFM. How many seconds would this take to fill a 50-gallon bag if you were to measure with the plastic bag method? ____

____

____

6) When performing the powered flow hood method, discuss the pressure inside the box and how it relates to the flow through the box. How do you know when to increase or decrease the speed of the fan? What are ultimately trying to achieve? ______________________________

Discussion and Critical Thinking:

1) Why is flow balancing useful? How would the diagnosis of poor or excessive airflow benefit the consumer? ______________________________

2) Discuss the pros and cons to each method of flow balancing found in this section ______________________________

3) Look into ACCA Manual J. How could the results of Manual J calculation be used to aid decision-making based on the results of flow balancing?

2.05 Combustion Appliance Zone (CAZ) Depressurization

Materials Needed:

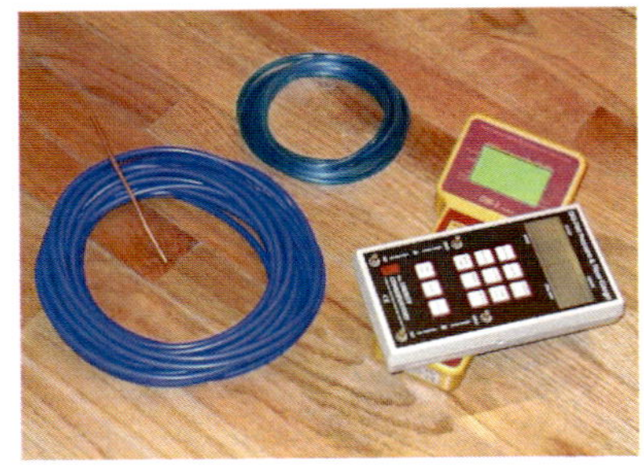

- digital manometer
- long and short pressure hoses
- 5/32" metal probe

CONCEPT

A Combustion Appliance Zone (CAZ) is any room that can be isolated from adjacent areas, and which contains combustion appliances. Your goal with this test is to make it as hard as possible for the combustion appliances in the tested CAZ to draft properly, using all the depressurization sources inherent in the home. Purposely do everything the homeowner might do accidentally that could force the home to suck on the combustion flues for makeup air.

All 4 of the following tests (*2.06-2.09*) take place under worst case conditions, as you can see in **Figure 2.05.1**, a chronological map of 5 Point Combustion Testing:

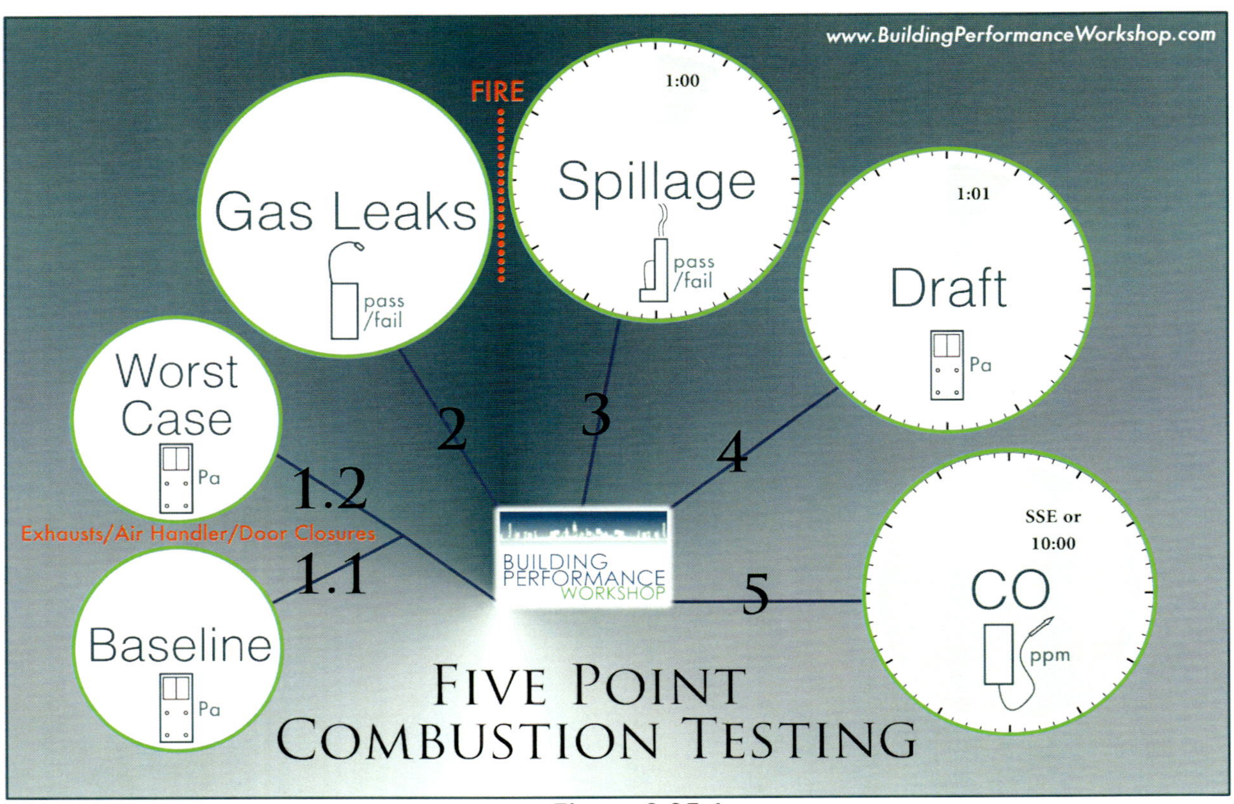

Figure 2.05.1

Based on the configuration of combustion appliances in the CAZ, BPI National Standards allow for depressurization up to, but not more negative than, the following Pascal limits (cited from BPI's Building Analyst Standard):

–2	Natural draft water heater alone (orphaned)
–3	Natural draft water heater commonly vented with natural draft boiler or furnace
–5	Natural draft boiler or furnace with damper commonly vented with water heater
–5	Individual natural draft boiler or furnace
–5	Induced draft boiler or furnace commonly vented with water heater
–15	Power vented or induced draft boiler or furnace alone, or fan assisted DHW alone
–50	Sealed combustion appliances, direct-vented appliances, chimney-top draft inducer, high static pressure flame retention head oil burner

Video demonstration available at www.BuildingPerformanceWorkshop.com/videos

PREPARATION

Put the home in baseline conditions according to A.05. Install one end of the long pressure hose outdoors by closing it in an exterior door (use the metal probe to keep from pinching the hose) or a small window, and take the hose's other end into the CAZ. ***TIP:** *Don't use a window that's in proximity to the CAZ, since your goal is to create a high pressure differential, and a slightly open window will relieve that pressure differential.*

Remove the forced air filter if it's dirty enough to restrict the system's airflow; otherwise, leave it installed as found.

TESTING

Turn on the manometer in Pr/Pr mode, and connect the reference tap of Channel A to the prepped pressure hose (though you can use either channel; *read A.03 for more*). Record the pressure reading on Channel A: this is your 'Baseline' pressure measurement. ***TIP:** *If the reading is fluctuating too much, increase the time average of the manometer.*

Turn on all exhaust devices in the home, including bath and kitchen fans, clothes dryers, etc. Do not activate whole house fans or the central air handler. Close the doors to any rooms that do not contain exhaust devices. ***TIP:** *If the home has a central vacuum system that exhausts into the garage, turn it on also, sucking air from the closest vacuum port to the CAZ.*

Return to the CAZ and record the pressure reading as the 'Exhaust Only' measurement. Disconnect the long pressure hose from the manometer, and connect the short pressure hose to an input tap, in order to measure the pressure in a zone with reference to the CAZ (*see A.03*).

Turn on the air handler (*see A.02*). If there are any supply registers in the CAZ, close them, but leave all other duct register settings exactly as found.

Go to the home's furthest room from the CAZ and toss into it the end of the short pressure hose. Close the door and read the pressure in the room with reference to the CAZ (since all the doors behind you are open, you're still technically standing in the CAZ). If the pressure in the room is positive, leave the door closed; if it's negative, leave the door open to provide a greater depressurization to the CAZ. ***TIP:** *Verify your manometer reading on the first door with chemical smoke to confirm you've set up the device correctly.*

Aside from your goal of depressurizing the CAZ, you also want to make the CAZ as small as possible to maximize the effect of the depressurization. Close the door to rooms that don't have duct registers or exhaust devices.

Work your way back to the CAZ, keeping it at your back always (to continue testing the pressure in the rooms with reference to the CAZ).

Once the home's interior doors are configured for maximum CAZ depressurization, unplug the short pressure hose from the manometer and reconnect the long pressure hose. Read the new pressure in the CAZ with reference to outdoors, and compare it to the 'Exhaust Only' measurement. If the new measurement is more negative than the 'Exhaust Only' measurement, continue to the next step. If the 'Exhaust Only' measurement was more negative than the new measurement, return the home to the 'Exhaust Only' state, and turn the air handler off.

Take the most negative pressure obtained as the 'Worst Case Depressurization' measurement. Subtract the 'Baseline' pressure measurement from this number to calculate the Adjusted Worst Case Depressurization. This number is the verified effect of the home's own depressurization sources upon the CAZ. ***TIP:** *You may find, due to stack effect or weather, that the 'Baseline' pressure measurement exceeds the CAZ depressurization limits before you've done anything to hurt the CAZ; this is common, and should not be considered an automatic failure of the test.*

Continue to measure the pressure in the CAZ with reference to outdoors throughout 2.06-2.09, in case anomalies are found in those tests that may be attributable to changes in the CAZ depressurization (obviously, gas leak detection will not be affected, but we train our students to this 5-Point Protocol for simplicity, and we use it ourselves at Chicago Home Performance for the same reason).

INTERPRETATION

If the CAZ depressurization exceeds the set limit, you should try to pinpoint which particular exhaust device is responsible. The kitchen exhaust hood is often the culprit; if so, test the effect of the "medium" and "low" settings to find out if the device should only be used on a lower setting.

QUIZ Combustion Appliance Zone (CAZ) Depressurization

1) Why would you want to use a window or door that is far from the CAZ to place your reference hose? ______________________________

____

____

2) Why would you want to close any supply registers that are located in the CAZ? ______________________________

____

3) Why is it important to shut any doors to rooms that do not have any supply registers, return registers, or exhaust fans? ______________________________

____

4) Why is it important to obtain an 'exhaust only' measurement prior to kicking on the air handler and performing door closures? ______________________________

____

____

5) A CAZ has the following characteristics:
   -Baseline = +0.3 Pa
   -Exhaust only = -3.4 Pa
   -Exhaust + Spacepak air handler and door closures = -4.4 Pa
   -Induced draft boiler commonly vented with ambient draft water heater.
   What is the worst-case depressurization? Does our CAZ pass BPI standards? ______________________________

____

____

6) A CAZ has the following characteristics:

-Baseline = -1.3 Pa

-Exhaust only = -3.1 Pa

-Exhaust + air handler and door closures = -2.9 Pa

-A power vented furnace with an orphaned ambient draft water heater.

What is the worst-case depressurization? Does our CAZ pass BPI standards? ______________________________

____

____

____

7) A CAZ has the following characteristics:

-Baseline = -0.9 Pa

-Exhaust only = -4.6 Pa

-Exhaust + air handler and door closures = -7.2 Pa

-A natural draft furnace commonly vented with a natural draft water heater.

What is the worst-case depressurization? Does our CAZ pass BPI standards? ______________________________

____

____

____

Discussion and Critical Thinking:

1) Think about the term 'worst case depressurization.' Why do you think this test is called this? ____

____

2) A bathroom with a properly ducted exhaust fan and a supply register is found to be positively pressurized. Even though the door will be closed, the exhaust fan should be left on. Why is this? ________________________

____

____

3) Discuss the difference between an 'orphaned' ambient-draft water heater and a stand-alone ambient-draft water heater. Research BPI technical standards and see how their standards address this issue. Can you think of why they classify these types separately and how it relates to worst-case depressurization? ______________________________________

____

____

____

4) Why is it a good idea to monitor the worst case CAZ depressurization throughout the entire series of test required for combustion safety testing? What will happen to the pressure in the CAZ as the combustion equipment kicks on? ____

____

____

____

2.06 Gas Leaks

Materials Needed:

- combustible gas detector
- bubble solution (soapy water)

PREPARATION

Turn off or pilot all combustion appliances according to A.02. Turn on the gas detector, or 'sniffer', and let it warm up. Adjust the device's sensitivity, to detect subtle increases in combustible gas. ***TIP:** *If you calibrate the sniffer outdoors, it may alert you to an elevated level of combustible gases in the home in general.*

Video demonstration available at www.BuildingPerformanceWorkshop.com/videos

TESTING

Sniff along the entirety of the accessible gas line, spending time at each joint, elbow, and valve. The detector 'sniffs' from its end, so make sure to keep it pointed at the gas line the entire time.

When the sniffer indicates that it's found a combustible gas (generally by buzzing or singing), verify that it is a gas leak by using bubble solution. It may not be a verifiable gas leak if something else in the vicinity is setting off the detector (gasoline, glue, some joint compounds, etc). ***HEADS UP:** *Natural gas is lighter*

than air, so concentrate on the top of those horizontal joints; propane is denser than air, so concentrate on the underside of those joints.

Watch for bubbles to form. They may form slowly, in a colony, or the bubble solution may get blown right off immediately if the leak is a big one. Be patient, and revisit the potential leakage site with both the gas sniffer (don't get it wet) and the bubble solution to be 100% certain of your diagnosis.

Use a mirror to see the far side of any hard-to-reach gas lines, in order to verify that bubbles are forming; if they are, tag the leak with a luggage tag or brightly colored tape. If the leak is inside the cabinet of a furnace or boiler, or anywhere near the burners, place the tag in a safe place and make sure a description of the exact location is written on the tag.

QUIZ Gas Leaks

1) Why is it a good idea to baseline the gas sniffer outdoors? ____________

2) What are the most important elements to check on the gas line? ____________

Discussion and Critical Thinking:

1) What do you think is an appropriate speed for checking for gas leaks? Is there a standard for this through any organization? ____________

2) If your gas sniffer indicates a leak but you cannot see bubbles form after soaping the area, what should you do? ____________

3) Why is it important to know what type of gas is in the line? How will this affect your testing? ____________

2.07 Combustion Spillage

Materials Needed:

- chemical smoke or mirror/sunglasses

CONCEPT

Between the combustion chamber of a natural draft appliance and the top of the chimney, there is always a draft control device: a draft diverter, hood, or barometric damper. When the combustion appliance is operating, air should be flowing into the exhaust flue through this device, never coming back out of it: that's called spillage. Generally, there is some degree of spillage when the appliance first fires, but it **must stop within 1 minute** according to BPI National Standards, under worst-case depressurization conditions (*see 2.05*). Pay special attention to flues that do not visibly pitch upward 1/4" per foot of run, or draft devices that show signs of moisture, rust, or mineral deposits.

A note on testing methods: the mirror or sunglasses will show spillage if they simply fog up, since one of the main ingredients in combustion gas is moisture. As far as the chemical smoke goes, there are a few choices:

- **Smoke pencil**
  - Pros: non-toxic, simple to use
  - Cons: buoyant by design, smoke odor, low visibility, low volume
- **Glycol Foggers**
  - Pros: non-toxic, heatless, quite visible, high volume, looks like a toy
  - Cons: battery operated, can be difficult to operate, looks like a toy
- **Titanium Tetrachloride**
  - Pros: heatless, quite visible, high volume
  - Cons: toxic if used improperly, cannot be stored near electronics
- **Powder**
  - Pros: non-toxic, simple to use
  - Cons: leaves residue, low visibility, low volume

Video demonstration available at www.BuildingPerformanceWorkshop.com/videos

PREPARATION

Turn off or pilot all combustion appliances according to A.02. Begin monitoring carbon monoxide (CO) levels in the air around you. **WARNING:** If at any point the CO level exceeds 35 ppm, stop testing, turn off the combustion appliance, open a window, and evacuate the home until the CO has returned to safe levels. Identify any draft devices or unintentional gaps/holes in the flue or chimney connection; these require testing.

TESTING

Fire the combustion appliance and start the one-minute countdown. Place the mirror/sunglasses or puff the smoke at the edge of the draft device. ***TIP:** *You want the device to show you spillage; if it spills in front of you, you can fix it or recommend a course of action. Therefore, don't aim your smoke into the flue; aim it parallel to the edge of the draft device and verify that it goes into the flue of its own volition.*

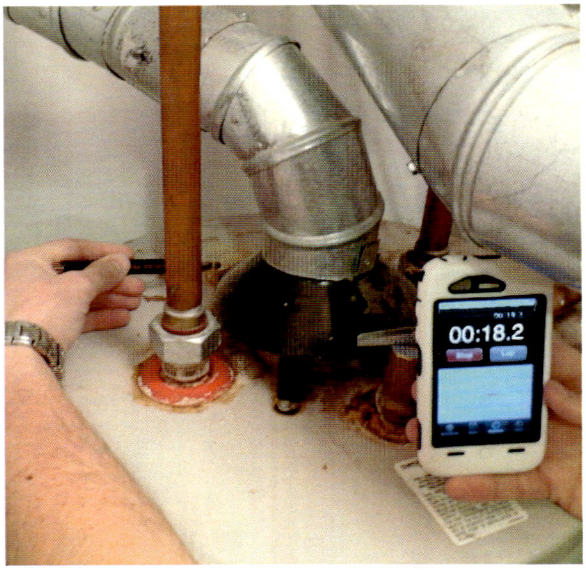

Check for spillage at every possible spillage location. This means 360° around draft diverters, or along the entirety of the draft hood's lip. Record the time it takes for the appliance to stop spilling (may be as early as 5 seconds, may take 59 seconds).

INTERPRETATION

If the appliance spills combustion gases after the one-minute time limit, there are a few possible causes.

Since you ran this test under worst-case depressurization conditions (*see 2.05*), you should first try to rule out the **CAZ depressurization** as the cause. You can perform the test again under natural, or 'Baseline' conditions (*see A.05*), after reconfiguring the home and waiting for the flue to cool down. Or, if you're clear-headed enough in the middle of testing, you can simply relieve the depressurization of the CAZ by opening wide the nearest window or door. If the spillage stops when the CAZ is at Baseline pressure, then you've pinpointed the cause.

Spillage may also be caused by **flue or chimney configuration problems or blockages**. If the draft pressure test (*see 2.08*) also fails AND the CAZ depressurization has been disproven as the cause, suspect a flue or chimney blockage. If the draft pressure is fine, that means at least some of the gases are getting successfully sucked out of the chimney, but not all of them are making it out. In this case, the cause may be under-sizing of the flue.

QUIZ Combustion Spillage

1) What could corrosion on the top of a natural drafting device possibly indicate? ______________________________

____

2) Why is it important to run this test under worst-case depressurization?

____

____

3) Why is it important to ensure that the flues are cool prior to testing for spillage?______________________________

____

4) What are the three likely causes of spillage? What are ways to manipulate your test settings to figure out the cause of spillage? ______________________________

____

____

Discussion and Critical Thinking:

1) What is the minimum rise per foot of run for a flue that is connected to natural drafting equipment? What is it about the properties of air at elevated temperatures that make this important? ______________________________

____

2) What is the difference between spillage and back drafting? ______________________________

____

3) Why do you think that a natural drafting device is likely to spill initially when firing after an extended dormant period? ______________________________

____

2.08 Draft Pressure

Materials Needed:

- digital manometer
- pressure hose
- 5/32" metal tube
- 12 or 18 volt drill and 5/32" drill bit

Video demonstration available at www.BuildingPerformanceWorkshop.com/videos

CONCEPT

Combustion appliances have two draft types: positive and negative. Natural draft and induced draft appliances are designed for negative draft in the flue, while power vented water heaters and sealed combustion appliances are designed for positive draft in the flue, from pushing the low temperature gases out of the home. Draft pressure testing under worst-case depressurization conditions (*see 2.05*) is standard for all negative draft appliances, and BPI National Standards are designed *only* for these.

Temperature outdoors	**Draft Limit (reading cannot be more positive than this)**
Below 10°F	–2.5 Pa
Between 10°F and 90°F	$Temp_{outdoors} \div 40 - 2.75$
Above 90°F	–0.5 Pa

A negative draft pressure in the flue with reference to the CAZ means the combustion gases are being sucked out of the chimney. A draft pressure that is at all positive is never good, and indicates a pushing of air back down the flue. Pay special attention to flues that do not visibly pitch upward 1/4" per foot of run.

PREPARATION

Begin monitoring carbon monoxide (CO) levels in the air around you. **WARNING:** If at any point the CO level exceeds 35 ppm, stop testing, turn off the combustion appliance, open a window, and evacuate the home until the CO has returned to safe levels.

Drill a 5/32" hole for the metal pressure probe in the flue of the natural draft or induced draft combustion appliance, 1-2 ft past the draft hood or first elbow. ***TIP:** *In an induced draft furnace, the draft inducer may create a slight positive pressure in the first section of the flue. Make sure you measure the draft after the first elbow to avoid any residual inducer pressure.*

Insert the metal probe, connected to one end of the pressure hose. Connect the other end of the hose to the input tap of either channel of the manometer (though you should always use the same channel, for consistency). I'll use Channel B for this demonstration.

TESTING

Fire the combustion appliance, and record the draft pressure on Channel B once the appliance passes or fails the spillage test (*see 2.07*). Since BPI national standards stipulate this test as 'Worst Case Draft', don't wait for the chimney to warm up. The hotter the chimney becomes, the better the draft, and your goal is to identify the weakest reading. This weakest draft pressure should exceed the draft limit and pass with flying colors.

INTERPRETATION

If the draft does not become more negative than the minimum limit, and it therefore fails the test, there are a few possible causes.

First, compare the draft test result with the **spillage test result** (*see 2.07*). They will generally agree with one another. If the spillage test passes but draft fails, the draft reading may be affected by turbulence in the flue. Try adjusting your flue probe to see if the reading changes.

Since you ran this test under worst-case depressurization conditions (*see 2.05*), you first should try to rule out the **CAZ depressurization** as the cause. You can perform the test again under natural, or 'Baseline' conditions (*see A.05*), after reconfiguring the home and waiting for the flue to cool down. Or, if you're clear-headed enough in the middle of testing, you can simply relieve the depressurization of the CAZ by opening wide the nearest window or door (if you're not clear-headed, you may enjoy sleep or prescribed medication). If the draft pressure improves significantly when the CAZ is at Baseline pressure, then you've pinpointed the cause.

QUIZ Draft Pressure

1) Worst Case CAZ depressurization is often confused with draft pressure. What are the fundamental differences between these measurements? How is the manometer configured in each test? ______________________________

____

____

____

2) Why would you want wait for a combustion appliance to pass spillage prior to recording worst-case draft pressure? How might a warm flue affect the draft pressure? ____

____

____

3) How might turbulent flow in the flue affect draft pressure readings? What are some ways to avoid the effects of turbulent flow? ____________________

____

____

4) On an 81° F degree day, the measured draft in an ambient draft water heater is found to be -1.2 Pa. Is this sufficient according to BPI national standards? ____

____

5) On an 8° F day, the measured draft on an induced draft furnace is -2.3 Pa. Is this sufficient according to BPI national standards? ___________________

____

Discussion and Critical Thinking:

1) Discuss the fundamental differences between positive and negative draft systems. Name several factors that have changed in furnace technology that would present a need for a positive draft system. ______________

____

____

____

2) Take a look at BPI national standards for Worst Case CAZ Depressurization and for Worst Case Draft Pressure. How do these two tests relate to the standards? How do the standards relate to mechanical system types? ______________________________

____

____

3) Discuss a scenario where you would want to test draft pressure when the CAZ is NOT in worst-case depressurization? What might you deduce from measuring draft pressure in 'natural conditions?' ______________

____

____

4) Discuss how Worst Case CAZ Depressurization, Spillage, and Worst Case Draft Pressure are all related. Can the results of any one test give an indication of how the other tests might go? Does it matter which order these tests are performed? __________________________

____

____

2.09 Combustion Analysis

Materials Needed:

- combustion analyzer (with temperature, carbon monoxide, and O_2 testing capability)
- 12 or 18 volt drill and 1/4" drill bit

Video demonstration available at www.BuildingPerformanceWorkshop.com/videos

CONCEPT

Running a combustion analysis is important for both safety and appliance efficiency. It would be great if every combustion appliance burned fuel perfectly, but they rarely do, and it's critical to know if there are safety or efficiency concerns. Carbon monoxide (CO) level in the flue gases is the most critical measurement for safety, but temperature, oxygen levels, and other readings can be useful in diagnosing a system that's not performing properly.

Many combustion testing standards stipulate 100 ppm as the maximum allowable CO level in the flue, though the BPI National Standards set the bar higher, with 25 ppm being the maximum allowable level. If the oxygen level is over 10%, the CO Air Free reading should be recorded, since there may be a serious issue that is masked by all the extra oxygen.

Most standards agree that over 400 ppm (at the low end of lethal levels) the combustion appliance should receive an emergency service, and you may decide to disable it until the service can be completed and the improved CO verified.

PREPARATION

Begin monitoring carbon monoxide levels in the air around you. **WARNING:** If at any point the CO level exceeds 35 ppm, stop testing, turn off the combustion appliance, open a window, and evacuate the home until the CO has returned to safe levels.

Identify the best location to sample combustion gases, before they mix with dilution air at a draft device (*see 2.07*). You may have to drill a hole in order to sample gases inside the throat of a water heater or before the draft hood of a boiler, for example.

Step outdoors to turn on and calibrate the combustion analyzer (unless otherwise directed by the manufacturer's instructions) to adjust for any elevated CO levels in the atmosphere at large. Then insert the analyzer's probe to be able to sample undiluted gases.

TESTING

Fire the combustion appliance, and monitor the carbon monoxide level as the appliance warms up. The CO level will peak in the first 30 seconds and taper off from there, with CO decreasing as the temperature increases. Wait until steady state is reached.

Once steady state is reached (when the analyzer's thermometer shows the flue temperature is no longer rising), record the temperature, carbon monoxide, and oxygen levels. ***HEADS UP:** *If your analyzer does not have a NOx filter installed, it may mistakenly give a higher than actual CO reading because of the NOx in the combustion gases.*

For sealed combustion or power vented appliances, insert the probe into the open end of the terminal at the side of the house, or drill into the PVC exhaust pipe according to A.04-C.

INTERPRETATION

Since you've performed this test in conjunction with 2.05 through 2.08, the most important result for this test is the carbon monoxide level at steady state. If the CO level is higher than the maximum allowable level, the appliance should be serviced with a clean-and-tune or a more intensive examination. The other readings, including temperature, oxygen, carbon dioxide, CO Air Free (what the CO would be if there were no oxygen left over from combustion), calculated efficiency, and so forth, can help you pinpoint the specific issue and the best way to fix it.

QUIZ Combustion Analysis

1) What is important about the testing location for CO? Specifically, what do you not want as part of the combustion gas sample? ____________________

____

____

2) Why is it important to wait until steady state is reached prior to testing for CO? What is typically a good indication that the appliance has reached steady state? ____________________

____

____

3) In regards to testing the CO in a flue, how does testing a sealed combustion appliance differ from testing a non-sealed combustion appliance? ____________________

____

____

4) Some modern combustion analyzers also measure draft pressure. What is important to remember about the testing location of CO and draft pressure? ____________________

____

____

Discussion and Critical Thinking:

1) Research what 'CO Air Free' means in relation to combustion analyzers. When and why would you want to use this type of measurement?

____

____

2) What does 'NOx' stand for? What is a 'NOx' filter? How would this filter alter the CO readings in the flue of a combustion appliance? ___________

____

____

3) Take a look at several sets of standards for CO produced from combustion appliances. Discuss any differences that you find from 'acceptable' levels. _______________________________________

____

____

4) Most combustion analyzers measure several operating parameters. What are some of these parameters and how can they be used to troubleshoot inefficient/unsafe operation of a combustion appliance?

____

____

____

____

2.10 Total Duct Leakage

Materials Needed:

- duct leakage testing fan
- digital manometer
- long and short pressure hoses
- duct mask tape and/or plastic wrap
- static pressure probe (optional)

Video demonstration available at www.BuildingPerformanceWorkshop.com/videos

CONCEPT

This test measures all the unintentional air leakage in the shell of a duct system. Some of the leakage will be to outdoors (a cost and waste issue), and some will be back into the home (a comfort and unbalancing issue). In order to discover how much of the leakage is going outdoors, you need to perform the Duct Leakage to Outside Test (*see 2.11*). In both cases, this is actually more a duct tightness test than a duct leakage test, since it induces a uniform pressure throughout the entire duct system, unlike the air handler, which creates pressures that diminish as distance increases.

Although you can test duct leakage at various pressures, I'll demonstrate here using the most widely used pressure of 25 Pa. This is equal to 0.1 inches of water column, a typical pressure inside the ductwork when the system is active.

PREPARATION

Turn off the air handler (*see A.02*) and put the home in baseline conditions according to A.05. Decide whether you'll be running a pressurization or depressurization test on the duct system. If any of the system's ductwork runs through an area where air quality is poor (for example, a damp crawlspace or dusty attic), then avoid the depressurization test, since it will bring air from that area into the home.

Determine where you'll install the duct leakage-testing fan. Get as close to the air handler as possible; mounting the fan directly to the air handler cabinet is possible by cutting a sheet of cardboard to fit. If the air handler cabinet is not accessible, use a central return.

Remove the air filter and seal the filter slot. Your goal is to test the entire duct system, and you've now removed the main blockage between the supply and return sides.

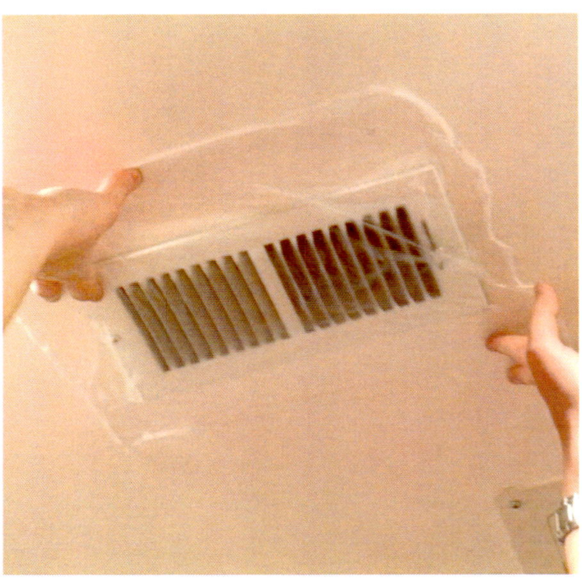

Without adjusting any manual dampers on duct registers, seal all intentional openings in the ductwork, including supply and return registers. Use slightly tacky plastic wrap only when performing a depressurization test; the test will pull the plastic firmly onto the register and ensure an airtight seal. ***TIP:** *Pay special attention*

to creating an airtight seal when registers are on a dusty or carpeted floor.

Connect the duct leakage-testing fan to the duct system at the chosen installation point (air handler cabinet or large central return). ***TIP:** *Attaching the fan directly, without the flex duct connector (or 'snorkel') increases the fan's airflow, which you may need if testing a leaky duct system.*

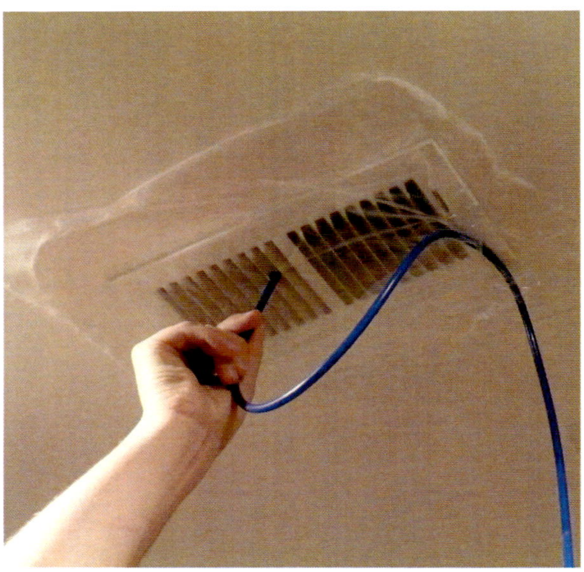

Insert one end of the long pressure hose into the duct system, at a register that's far from the duct-testing fan. You can insert it at any supply register, or in the supply trunk or plenum, but always keep the air handler between the testing fan and the pressure probe. The concept here is the same as in the blower door test (*see 1.01*): to protect the end of the pressure hose from the fan's turbulence. If the duct-testing fan might create turbulence at your sampling location, use the static pressure probe instead of the bare hose. ***TIP:** *In very leaky duct systems, flow readings may vary based on the probe's location; in the case of a high flow reading, sample at a variety of proximities to the testing fan.*

Configure the testing fan's digital manometer, and connect the pressure hoses so that Channel A is measuring the pressure in the duct system with reference to the home, and Channel B is measuring the flow in the fan with reference to the pre-fan zone (*for more information, see A.03, 'Manometer Mechanics'*).

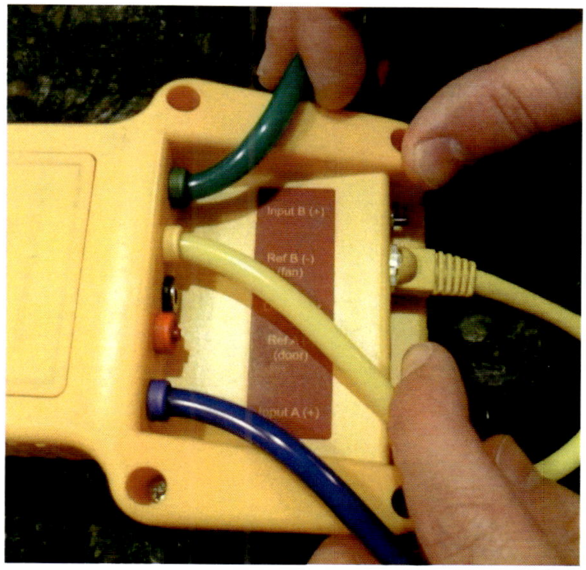

If the testing fan is installed in unconditioned space, or if a significant portion of a potentially leaky duct system is located in unconditioned space, open a window between the home and outdoors. This will prevent pressurizing or depressurizing the home (which is a crucial reference for Channel A) with the testing fan.

TESTING

Turn on the testing fan and bring the pressure in the ducts to (+/–) 25 Pa with reference to the home. Record the flow on Channel B. If you have trouble reaching 25 Pa (which is common), shut down the testing fan and do a sweep through the home and verify that all registers are sealed. Confirm that the hoses, manometer, and testing fan are

configured correctly, and run the test again. ***TIP:** *To ensure the equipment is configured correctly, measure the flow at the highest pressure the fan can reach, then measure the flow at exactly half that pressure. The airflow at half pressure should be approximately 0.65 times the airflow at the highest pressure.*

To determine the proportion of supply side leakage to return side leakage, reinstall the filter wrapped in plastic, reseal the filter slot, and re-run the test. Now you have effectively blocked the connection of the return and supply sides.

INTERPRETATION

There are a myriad of ways to interpret your test data for your client:

Percentage of Total System Airflow:

Duct Leakage as a Percentage of Total System Airflow is calculated by dividing the duct leakage rate by the air handler flow rate. Air handler flow rates can be measured or estimated using any of the techniques demonstrated in 2.01. This interpretation may be misleading, since your test results at a uniform 25 Pa throughout the ductwork will generally overestimate the system's leakage during normal operation.

California's Title 24 Energy Code, for example, requires that all new houses have a measured total duct leakage rate of less than 6% of total system airflow.

Equation 2.10.1 $$\%_{Airflow} = \frac{Q_{25}}{Q_{total}} \times 100$$

Where:

Q_{25} = Total duct leakage measured at 25 Pa [CFM]

Q_{total} = Total system airflow (*from 2.01*) [CFM]

Example:

Q_{25} = 362 CFM at 25 Pa

Q_{total} = 1242 CFM total system airflow

$$\%_{Airflow} = \frac{362}{1242} \times 100 = 29\%$$

Duct Leakage as a Percent of Floor Area:

The International Energy Conservation Code®, ENERGY STAR® Homes, and other standards define reasonable duct leakage according to the floor area served by the duct system. In **Equation 2.10.2**, percentage of total floor area is the same as CFM per 100 ft^2 of floor area, which is how it's often phrased.

Equation 2.10.2 $$\%_{Floor} = \frac{Q_{25}}{A_{Served}} \times 100$$

Where:

Q_{25} = Total duct leakage measured at 25 Pa [CFM]

A_{Served} = Floor area served by duct system [ft^2]

Example:

Q_{25} = 362 CFM at 25 Pa

A_{Served} = 2000 ft^2

$$\%_{Floor} = \frac{362}{2000} \times 100 = 18\% \textit{ (or 18 CFM per 100 ft}^2\textit{)}$$

Estimated Duct Leakage Areas:

Your client may better understand a Leakage Area of the duct system in square inches, which is a useful way to visualize the physical size of all cumulative leaks in the duct system. Using a test pressure of 25 Pa, **Equation 2.10.3** will allow you to calculate the Equivalent Orifice Leakage Area (EOLA):

Equation 2.10.3 $$A_{Leakage} = \frac{Q_{25}}{5.3}$$

Where:

$A_{Leakage}$ = Cumulative area of all leaks [in^2]

Q_{25} = Total duct leakage measured at 25 Pa [CFM]

Example:

Q_{25} = 362 CFM at 25 Pa

$$A_{Leakage} = \frac{362}{5.3} = \mathbf{68\ in^2\ EOLA}$$

QUIZ Total Duct Leakage

1) Why is it desirable to hook up the fan directly to the blower cabinet without the flex duct connector (the 'snorkel')? ____________________

____

____

2) Why is it important to open a window or door when testing a duct system that has a significant portion of the ducts located outside the conditioned space? How might this affect the results? ____________________

____

____

____

3) In a home that is required to have a duct leakage rate of 10% or less of total system airflow, you measure airflow through the duct system in cooling mode with a Pitot tube and find a flow of 2980 CFM. A total duct leakage test revealed a leakage rate of 242 CFM_{25}. Does this meet the requirement? What is the actual leakage expressed as a percentage?

____

____

____

____

4) What is the EOLA of the duct system from question 3? ____________________

____

____

5) A building inspector calls your office to verify that a home that your colleague tested that is in his jurisdiction has passed the 2009 IECC tightness limits of 12 CFM per 100 ft^2. You review your files, and discover

that the 4320 ft^2 home did indeed pass with a total duct leakage of 9.3 CFM per 100 ft^2, but that is all of the information you have. What was the flow through the fan that your colleague tested? ______________________

____

____

____

____

Discussion and Critical Thinking:

1) Can you think of reasons why it is best to have the duct-testing fan located as close to the air handler as possible? ____________________

____

____

____

2) What are a few reasons you would not want to depressurize the duct system, but instead would want to pressurize the system? ___________

____

____

____

3) The manometer set up for this test is quite complicated. Review **A.03** and then discuss in detail how the manometer is set up for this test.

____

____

____

____

4) If a home has multiple duct systems serving different zones of the home, how do these methods of quantifying duct leakage apply? Is it

necessary to break out each system to a specific area or test all systems and then have cumulative leakage rate for the entire area of the home? What if some systems are located outside the envelope and some are inside the envelope and local codes mandate that only ducts outside the envelope must be tested? Discuss the different ways to interpret these scenarios. ____

____

____

____

____

____

5) If all ductwork is clearly located inside the envelope, is duct leakage testing a valuable measurement? Why or why not? ____________________

____

____

____

____

6) Research various standards or codes and how they relate to duct tightness limits. Which have you found to be the strictest? Are there any codes or standards that exempt this test if certain conditions are met in the home? When would this be the case? ______________________________

____

____

____

____

2.11 Duct Leakage to Outside

Materials Needed:

- duct leakage testing fan
- digital manometer
- pressure hoses
- duct mask tape and/or plastic wrap
- blower door (complete kit)

Video demonstration available at www.BuildingPerformanceWorkshop.com/videos

CONCEPT

This test measures the unintentional air leakage between the duct system and outdoors, which is both a comfort and an overall waste issue. By pitting the blower door and the duct-testing fan against each other, you can force both of them to go outside their relationship for air makeup or exhaust needs.

PREPARATION

Determine whether you'll be using pressurization or depressurization, and set up the total duct leakage test according to 2.10.

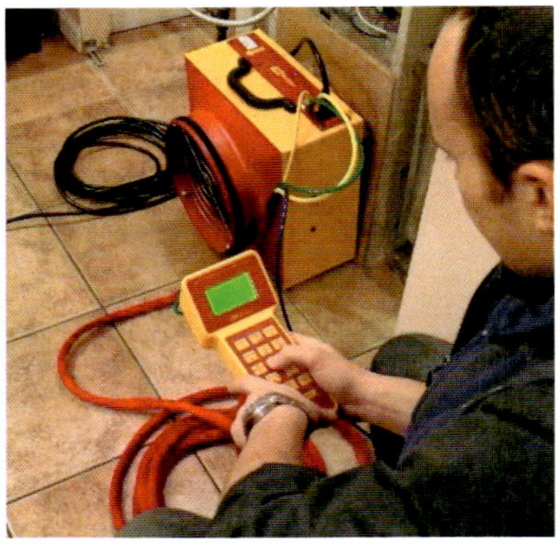

Set up the blower door test according to 1.01, up to and including baselining the manometer. Use the same airflow direction as the duct-testing fan (pressurization or depressurization).

TESTING

Run the blower door up until the home reaches +/- 25 Pa with reference to outdoors, using manometer automation (cruise control) or a helper; the home pressure will start to slide once the duct leakage testing begins.

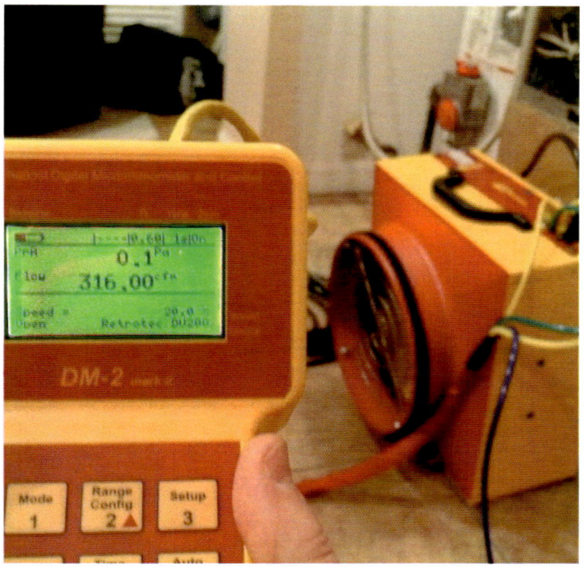

Run the duct leakage testing fan up until the pressure in the duct system is **0 Pa** with reference to the home. At that point, all the airflow across both the duct system and the envelope is to/from outdoors.

INTERPRETATION

Analyze this test data using the same equations as in 2.10.

QUIZ Duct Leakage to Outside

1) In order for a fluid to move, two conditions must be met: a pressure differential between two zones and a pathway to connect them. Describe in detail how this test manipulates the pressure differential to control where the pathway is located. ______________________________

____

____

____

____

2) Why would it always be more beneficial to use manometer automation ('cruise control') for the blower door instead of manually driving the house to a specified pressure during duct leakage to outside testing? ____________

____

____

____

3) Once the blower door is running and before the duct testing fan is turned on, what would you expect to read on the pressure channel of the manometer for a leaky duct system? What about a tight duct system? What does it mean to 'run the duct leakage testing fan up until the pressure in the duct system is 0 Pa WRT the home?' ______________________________

____

____

____

____

4) A 6200 ft^2 home has 3 duct systems, 2 of which are located in the unconditioned attic that serve the 2800 ft^2 top floor. The other duct system is clearly located in conditioned space. The local energy code only requires

ductwork located outside the envelope to be tested for leakage to outdoors and it must meet a requirement of 6 CFM per 100 ft^2 served. One of the top floor systems has 72 CFM of leakage to outdoors and the other has 91 CFM. Do these duct systems meet this requirement? What is the actual leakage in CFM per 100 ft^2 served? ______________________________

5) If the total leakage for the ducts in problem 4 were 145 CFM and 167 CFM respectively, what is the CFM per 100 ft^2 for total leakage? What is the total EOLA for the duct systems? What is the theoretical EOLA to outdoors? What is the theoretical EOLA to indoors? ______________________________

Discussion and Critical Thinking:

1) Discuss the fundamental difference between total duct leakage and duct leakage to outside. ______________________________

2) Why is it important to have the blower door and the duct blaster in the same mode (pressurization or depressurization)? What would happen if the were in different modes? ______________________________

3) A duct system is found to have a very high total leakage rate but a very low leakage to outside rate, what sort of issues does this present for the homeowner? What would the implications be if the leakage to outside rate were also very high?

4) Energy codes generally require this test prior to occupancy permits are issued. Some codes have a clause that allows for this test to be performed at 'rough in' as opposed to the final stages of construction. List some reasons why a builder would want to perform this test at 'rough in.'

5) In a conversation with an architect, he/she claims he doesn't want to place ductwork inside the envelope for aesthetic reasons. Using what you know about this test, list a few reasons why you would convince him/her otherwise.

2.12 Exhaust Fan Flow

2.12-A Flow Pan Method

Materials Needed:

- flow pan (such as Energy Conservatory's Exhaust Fan Flow Meter)
- digital manometer
- short pressure hose

Video demonstration available at www.BuildingPerformanceWorkshop.com/videos

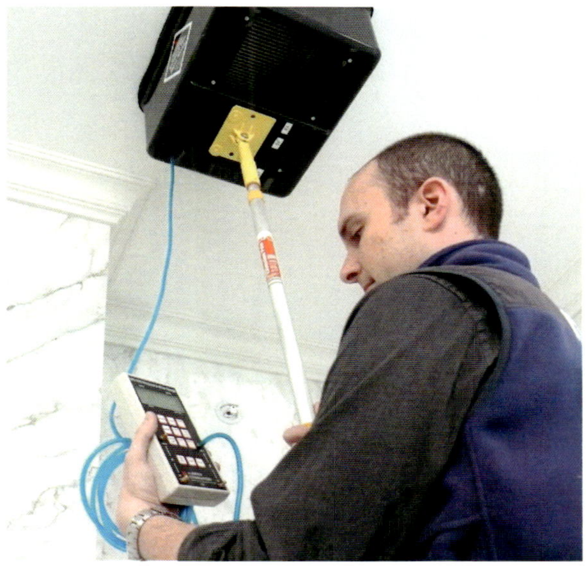

PREPARATION

Using the pressure hose, connect the tap on the flow pan to the input tap on Channel B of the manometer. Start with the most wide-open hole configuration.

TESTING

Turn on the exhaust fan. Cover the entire fan with the flow pan. Record the airflow.

2.12-B Flow Hood Method

Materials Needed:

- flow hood (also called a balometer)

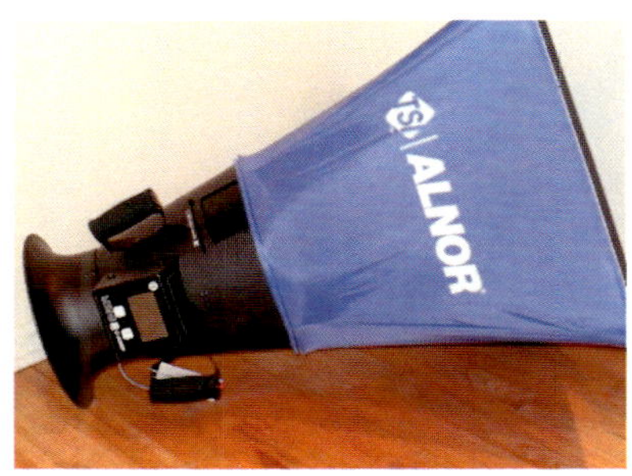

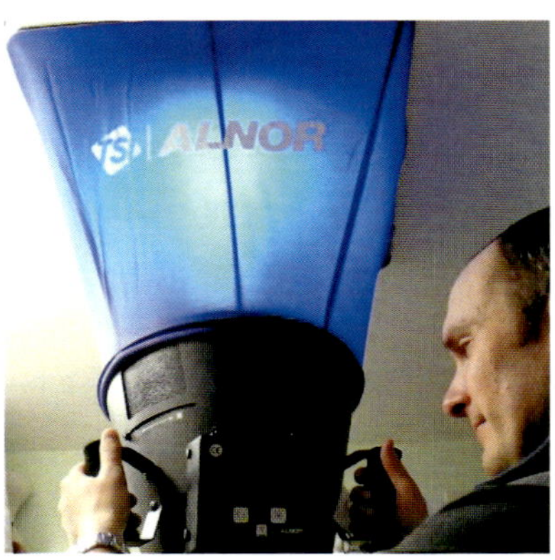

TESTING

Turn on the exhaust fan and cover it completely with the balometer. Record the indicated airflow.

QUIZ Exhaust Fan Flow

1) What would happen to the flow reading if the flow pan or hood could not cover the entire exhaust fan? Can you think of any situations where this would be the case? ______________________________

____

____

Discussion and Critical Thinking:

1) Is the pressure in the box positive or negative when measuring exhaust fan flow? What would happen if you put the box over a supply register? Is a flow pan designed to be used this way? Is a flow hood designed to be used this way? ______________________________

____

____

2) The Exhaust Fan Flow Meter from The Energy Conservatory has a published table that tabulates the flow through the pan at various pressures found in the pan WRT ambient conditions. Find this table and discuss the range of pressures and flows listed. What do you notice about the configuration of the device? Discuss how the rings of a blower door are similar to the configuration of the Exhaust Fan Flow Meter.

____

____

____

____

3) Why are these tests relevant? What are some situations where you would want to know the exhaust fan flow? ______________________________

____

____

2.13 Superheat and Subcooling

CONCEPT

Air conditioner and heat pump systems have two elements: airflow and refrigerant. It's important that both of these elements are performing optimally, and the performance of the refrigerant is tested by measuring the superheat and subcooling. **HEADS UP:** *Before running this test, make sure you have verified the Total System Airflow according to 2.01, and cleaned both coils; any problems found in the refrigerant loop may be due to airflow problems or dirt/dust insulating the coils, so save yourself the trouble and make sure these things are perfect first.*

There are a few different types of refrigerant you'll find in a residential system (mostly R-22 and R-410A). All refrigerants have a 'saturation temperature', which is the same thing as a boiling/condensing point (more precisely, it's the highest saturation of heat for a liquid not to boil, or the lowest saturation of heat for a vapor not to condense). The higher the pressure, the higher the saturation temperature; most refrigerants boil at –40°F at normal atmospheric pressure, which means it can hurt you, so personal protective equipment is important for 2.13-B.

Superheat is the temperature added on top of the saturation temperature, after the refrigerant has absorbed heat from the home into the suction line. Subcooling is the temperature below the saturation temperature, after the refrigerant has lost much of its heat to the outdoors from the liquid line.

There are in fact two methods for testing, though the first is commonly overlooked. 2.13-A should always be used first, so the integrity of the sealed system is upheld unless absolutely necessary. There is no reason for a sealed system to lose charge (unless, of course, you count the day once a year when a technician pumps a little of it into a set of hoses and gauges and leaves with it).

2.13-A Non-Invasive, Gauge-less Method

2.13-B Pressure Gauge Method

Figure 2.13.1

Target Total Superheat Table (non-TXV systems only)

TARGET EVAPORATOR SUPERHEAT FOR TXV-METERED SYSTEMS:

6-12ºF

If the equipment manufacturer's charging chart is available, always defer to that chart instead.

excerpted from California's 2008 Title 24 Residential Compliance Manual, with permission from the California Energy Commission ©2008

Return Air Wet-Bulb Temperature [°F]	Condenser Air Dry-Bulb Temperature [°F]																									
	90	**91**	**92**	**93**	**94**	**95**	**96**	**97**	**98**	**99**	**100**	**101**	**102**	**103**	**104**	**105**	**106**	**107**	**108**	**109**	**110**	**111**	**112**	**113**	**114**	**115**
50	5.0	5.0	5.0	5.0	5.0	5.0	5.0	5.0	5.0	5.0	5.0	5.0	5.0	5.0	5.0	5.0	5.0	5.0	5.0	5.0	5.0	5.0	5.0	5.0	5.0	5.0
51	5.0	5.0	5.0	5.0	5.0	5.0	5.0	5.0	5.0	5.0	5.0	5.0	5.0	5.0	5.0	5.0	5.0	5.0	5.0	5.0	5.0	5.0	5.0	5.0	5.0	5.0
52	5.0	5.0	5.0	5.0	5.0	5.0	5.0	5.0	5.0	5.0	5.0	5.0	5.0	5.0	5.0	5.0	5.0	5.0	5.0	5.0	5.0	5.0	5.0	5.0	5.0	5.0
53	5.0	5.0	5.0	5.0	5.0	5.0	5.0	5.0	5.0	5.0	5.0	5.0	5.0	5.0	5.0	5.0	5.0	5.0	5.0	5.0	5.0	5.0	5.0	5.0	5.0	5.0
54	5.0	5.0	5.0	5.0	5.0	5.0	5.0	5.0	5.0	5.0	5.0	5.0	5.0	5.0	5.0	5.0	5.0	5.0	5.0	5.0	5.0	5.0	5.0	5.0	5.0	5.0
55	5.0	5.0	5.0	5.0	5.0	5.0	5.0	5.0	5.0	5.0	5.0	5.0	5.0	5.0	5.0	5.0	5.0	5.0	5.0	5.0	5.0	5.0	5.0	5.0	5.0	5.0
56	5.0	5.0	5.0	5.0	5.0	5.0	5.0	5.0	5.0	5.0	5.0	5.0	5.0	5.0	5.0	5.0	5.0	5.0	5.0	5.0	5.0	5.0	5.0	5.0	5.0	5.0
57	5.0	5.0	5.0	5.0	5.0	5.0	5.0	5.0	5.0	5.0	5.0	5.0	5.0	5.0	5.0	5.0	5.0	5.0	5.0	5.0	5.0	5.0	5.0	5.0	5.0	5.0
58	5.0	5.0	5.0	5.0	5.0	5.0	5.0	5.0	5.0	5.0	5.0	5.0	5.0	5.0	5.0	5.0	5.0	5.0	5.0	5.0	5.0	5.0	5.0	5.0	5.0	5.0
59	5.0	5.0	5.0	5.0	5.0	5.0	5.0	5.0	5.0	5.0	5.0	5.0	5.0	5.0	5.0	5.0	5.0	5.0	5.0	5.0	5.0	5.0	5.0	5.0	5.0	5.0
60	5.0	5.0	5.0	5.0	5.0	5.0	5.0	5.0	5.0	5.0	5.0	5.0	5.0	5.0	5.0	5.0	5.0	5.0	5.0	5.0	5.0	5.0	5.0	5.0	5.0	5.0
61	5.0	5.0	5.0	5.0	5.0	5.0	5.0	5.0	5.0	5.0	5.0	5.0	5.0	5.0	5.0	5.0	5.0	5.0	5.0	5.0	5.0	5.0	5.0	5.0	5.0	5.0
62	5.0	5.0	5.0	5.0	5.0	5.0	5.0	5.0	5.0	5.0	5.0	5.0	5.0	5.0	5.0	5.0	5.0	5.0	5.0	5.0	5.0	5.0	5.0	5.0	5.0	5.0
63	6.8	6.1	5.4	5.0	5.0	5.0	5.0	5.0	5.0	5.0	5.0	5.0	5.0	5.0	5.0	5.0	5.0	5.0	5.0	5.0	5.0	5.0	5.0	5.0	5.0	5.0
64	8.8	8.1	7.5	6.8	6.2	5.6	5.0	5.0	5.0	5.0	5.0	5.0	5.0	5.0	5.0	5.0	5.0	5.0	5.0	5.0	5.0	5.0	5.0	5.0	5.0	5.0
65	10.9	10.3	9.8	9.2	8.7	8.1	7.5	7.0	6.4	5.8	5.3	5.0	5.0	5.0	5.0	5.0	5.0	5.0	5.0	5.0	5.0	5.0	5.0	5.0	5.0	5.0
66	12.8	12.2	11.7	11.1	10.6	10.0	9.5	8.9	8.4	7.9	7.3	6.8	6.2	5.7	5.2	5.0	5.0	5.0	5.0	5.0	5.0	5.0	5.0	5.0	5.0	5.0
67	14.6	14.1	13.5	13.0	12.5	12.0	11.4	10.9	10.4	9.9	9.3	8.8	8.3	7.8	7.2	6.7	6.2	5.7	5.2	5.0	5.0	5.0	5.0	5.0	5.0	5.0
68	16.5	15.9	15.4	14.9	14.4	13.9	13.4	12.9	12.4	11.9	11.4	10.9	10.4	9.9	9.3	8.8	8.3	7.9	7.4	6.9	6.4	5.9	5.4	5.0	5.0	5.0
69	18.3	17.8	17.3	16.8	16.3	15.8	15.3	14.9	14.4	13.9	13.4	12.9	12.4	11.9	11.5	11.0	10.5	10.0	9.5	9.1	8.6	8.1	7.6	7.2	6.7	6.2
70	20.1	19.7	19.2	18.7	18.2	17.8	17.3	16.8	16.4	15.9	15.4	15.0	14.5	14.0	13.6	13.1	12.6	12.2	11.7	11.3	10.8	10.4	9.9	9.5	9.0	8.6
71	22.0	21.5	21.1	20.6	20.2	19.7	19.2	18.8	18.3	17.9	17.5	17.0	16.6	16.1	15.7	15.2	14.8	14.4	13.9	13.5	13.1	12.6	12.2	11.8	11.4	10.9
72	23.8	23.4	22.9	22.5	22.1	21.6	21.2	20.8	20.3	19.9	19.5	19.1	18.6	18.2	17.8	17.4	17.0	16.6	16.1	15.7	15.3	14.9	14.5	14.1	13.7	13.3
73	25.6	25.2	24.8	24.4	24.0	23.6	23.2	22.7	22.3	21.9	21.5	21.1	20.7	20.3	19.9	19.5	19.1	18.7	18.4	18.0	17.6	17.2	16.8	16.4	16.1	15.7
74	27.5	27.1	26.7	26.3	25.9	25.5	25.1	24.7	24.3	24.0	23.6	23.2	22.8	22.4	22.1	21.7	21.3	21.0	20.6	20.2	19.9	19.5	19.1	18.8	18.4	18.1
75	29.3	28.9	28.5	28.2	27.8	27.4	27.1	26.7	26.3	26.0	25.6	25.3	24.9	24.5	24.2	23.8	23.5	23.2	22.8	22.5	22.1	21.8	21.5	21.1	20.8	20.5
76	31.1	30.8	30.4	30.1	29.7	29.4	29.0	28.7	28.3	28.0	27.7	27.3	27.0	26.7	26.3	26.0	25.7	25.4	25.1	24.7	24.4	24.1	23.8	23.5	23.2	22.9

Condenser Air Dry-Bulb Temperature [°F] \ Return Air Wet-Bulb Temperature [°F]	50	51	52	53	54	55	56	57	58	59	60	61	62	63	64	65	66	67	68	69	70	71	72	73	74	75	76
60	7.0	8.4	9.8	11.2	12.6	14.0	15.4	16.8	18.2	19.6	21.0	22.4	23.8	25.2	26.6	28.1	29.6	31.0	32.4	33.7	35.1	36.4	37.8	39.1	40.4	41.6	42.9
61	6.5	7.9	9.3	10.7	12.1	13.5	14.9	16.3	17.7	19.1	20.5	21.9	23.3	24.7	26.1	27.5	29.0	30.4	31.8	33.2	34.6	35.9	37.3	38.6	39.9	41.2	42.4
62	6.0	7.4	8.8	10.2	11.7	13.1	14.5	15.9	17.3	18.7	20.1	21.4	22.8	24.2	25.5	27.0	28.4	29.9	31.3	32.7	34.1	35.4	36.8	38.1	39.4	40.7	42.0
63	5.3	6.8	8.3	9.7	11.1	12.6	14.0	15.4	16.8	18.2	19.6	20.9	22.3	23.6	25.0	26.4	27.8	29.3	30.7	32.2	33.6	34.9	36.3	37.7	39.0	40.3	41.6
64	5.0	6.1	7.6	9.1	10.6	12.0	13.5	14.9	16.3	17.7	19.0	20.4	21.7	23.1	24.4	25.8	27.3	28.7	30.2	31.6	33.0	34.4	35.8	37.2	38.5	39.9	41.2
65	5.0	5.4	7.0	8.5	10.0	11.5	12.9	14.3	15.8	17.1	18.5	19.9	21.2	22.5	23.8	25.2	26.7	28.2	29.7	31.1	32.5	33.9	35.3	36.7	38.1	39.4	40.8
66	5.0	5.0	6.3	7.8	9.3	10.8	12.3	13.8	15.2	16.6	18.0	19.3	20.7	22.0	23.2	24.6	26.1	27.6	29.1	30.6	32.0	33.4	34.9	36.3	37.6	39.0	40.4
67	5.0	5.0	5.5	7.1	8.7	10.2	11.7	13.2	14.6	16.0	17.4	18.8	20.1	21.4	22.7	24.1	25.6	27.1	28.6	30.1	31.5	33.0	34.4	35.8	37.2	38.6	39.9
68	5.0	5.0	5.0	6.3	8.0	9.5	11.1	12.6	14.0	15.5	16.8	18.2	19.5	20.8	22.1	23.5	25.0	26.5	28.0	29.5	31.0	32.5	33.9	35.3	36.8	38.1	39.5
69	5.0	5.0	5.0	5.5	7.2	8.8	10.4	11.9	13.4	14.8	16.3	17.6	19.0	20.3	21.5	22.9	24.4	26.0	27.5	29.0	30.5	32.0	33.4	34.9	36.3	37.7	39.1
70	5.0	5.0	5.0	5.0	6.4	8.1	9.7	11.2	12.7	14.2	15.7	17.0	18.4	19.7	20.9	22.3	23.9	25.4	27.0	28.5	30.0	31.5	33.0	34.4	35.9	37.3	38.7
71	5.0	5.0	5.0	5.0	5.6	7.3	8.9	10.5	12.1	13.6	15.0	16.4	17.8	19.1	20.3	21.7	23.3	24.9	26.4	28.0	29.5	31.0	32.5	34.0	35.4	36.9	38.3
72	5.0	5.0	5.0	5.0	5.0	6.4	8.1	9.8	11.4	12.9	14.4	15.8	17.2	18.5	19.7	21.2	22.8	24.3	25.9	27.4	29.0	30.5	32.0	33.5	35.0	36.5	37.9
73	5.0	5.0	5.0	5.0	5.0	5.6	7.3	9.0	10.7	12.2	13.7	15.2	16.6	17.9	19.2	20.6	22.2	23.8	25.4	26.9	28.5	30.0	31.5	33.1	34.6	36.0	37.5
74	5.0	5.0	5.0	5.0	5.0	5.0	6.5	8.2	9.9	11.5	13.1	14.5	15.9	17.3	18.6	20.0	21.6	23.2	24.8	26.4	28.0	29.5	31.1	32.6	34.1	35.6	37.1
75	5.0	5.0	5.0	5.0	5.0	5.0	5.6	7.4	9.2	10.8	12.4	13.9	15.3	16.7	18.0	19.4	21.1	22.7	24.3	25.9	27.5	29.1	30.6	32.2	33.7	35.2	36.7
76	5.0	5.0	5.0	5.0	5.0	5.0	5.0	6.6	8.4	10.1	11.7	13.2	14.7	16.1	17.4	18.9	20.5	22.1	23.8	25.4	27.0	28.6	30.1	31.7	33.3	34.8	36.3
77	5.0	5.0	5.0	5.0	5.0	5.0	5.0	5.7	7.5	9.3	11.0	12.5	14.0	15.4	16.8	18.3	20.0	21.6	23.2	24.9	26.5	28.1	29.7	31.3	32.8	34.4	36.0
78	5.0	5.0	5.0	5.0	5.0	5.0	5.0	5.0	6.7	8.5	10.2	11.8	13.4	14.8	16.2	17.7	19.4	21.1	22.7	24.4	26.0	27.6	29.2	30.8	32.4	34.0	35.6
79	5.0	5.0	5.0	5.0	5.0	5.0	5.0	5.0	5.9	7.7	9.5	11.1	12.7	14.2	15.6	17.1	18.8	20.5	22.2	23.8	25.5	27.1	28.8	30.4	32.0	33.6	35.2
80	5.0	5.0	5.0	5.0	5.0	5.0	5.0	5.0	5.0	6.9	8.7	10.4	12.0	13.5	15.0	16.6	18.3	20.0	21.7	23.3	25.0	26.7	28.3	29.9	31.6	33.2	34.8
81	5.0	5.0	5.0	5.0	5.0	5.0	5.0	5.0	5.0	6.0	7.9	9.7	11.3	12.9	14.3	16.0	17.7	19.4	21.1	22.8	24.5	26.2	27.9	29.5	31.2	32.8	34.4
82	5.0	5.0	5.0	5.0	5.0	5.0	5.0	5.0	5.0	5.2	7.1	8.9	10.6	12.2	13.7	15.4	17.2	18.9	20.6	22.3	24.0	25.7	27.4	29.1	30.7	32.4	34.0
83	5.0	5.0	5.0	5.0	5.0	5.0	5.0	5.0	5.0	5.0	6.3	8.2	9.9	11.6	13.1	14.9	16.6	18.4	20.1	21.8	23.5	25.2	26.9	28.6	30.3	32.0	33.7
84	5.0	5.0	5.0	5.0	5.0	5.0	5.0	5.0	5.0	5.0	5.5	7.4	9.2	10.9	12.5	14.3	16.1	17.8	19.6	21.3	23.0	24.8	26.5	28.2	29.9	31.6	33.3
85	5.0	5.0	5.0	5.0	5.0	5.0	5.0	5.0	5.0	5.0	5.0	6.6	8.5	10.3	11.9	13.7	15.5	17.3	19.0	20.8	22.6	24.3	26.0	27.8	29.5	31.2	32.9
86	5.0	5.0	5.0	5.0	5.0	5.0	5.0	5.0	5.0	5.0	5.0	5.8	7.8	9.6	11.3	13.2	15.0	16.7	18.5	20.3	22.1	23.8	25.6	27.3	29.1	30.8	32.6
87	5.0	5.0	5.0	5.0	5.0	5.0	5.0	5.0	5.0	5.0	5.0	5.0	7.0	8.9	10.6	12.6	14.4	16.2	18.0	19.8	21.6	23.4	25.1	26.9	28.7	30.4	32.2
88	5.0	5.0	5.0	5.0	5.0	5.0	5.0	5.0	5.0	5.0	5.0	5.0	6.3	8.2	10.0	12.0	13.9	15.7	17.5	19.3	21.1	22.9	24.7	26.5	28.3	30.1	31.8
89	5.0	5.0	5.0	5.0	5.0	5.0	5.0	5.0	5.0	5.0	5.0	5.0	5.5	7.5	9.4	11.5	13.3	15.1	17.0	18.8	20.6	22.4	24.3	26.1	27.9	29.7	31.5

Shaded area requires return plenum temperature of 70°F or higher.

2.13-A Non-Invasive, Gauge-less Method

Materials Needed:

- wet bulb/dry bulb probe thermometer
- clamp or contact thermocouple (thermometer)

CONCEPT

Since you've already cleaned the coils and run 2.01 to verify the total system airflow is correct (~400 CFM per ton of cooling), if there is a problem with the air conditioning system's temperatures, you can deduce that it must be a refrigerant problem. Therefore, in this test method, you will assume the refrigerant is operating at optimal pressures, and no pressure gauges are needed. If you successfully identify a refrigerant issue with this non-invasive method, the pressure gauges won't tell you anything you don't already know, unless you're an HVAC specialist and understand the intricacies of the system's design. Using digital tools will give you an accuracy of +/–3°F, better than the analog accuracy of +/–5°F. I'm grateful to Jim Bergmann of TruTech Tools for having perfected and published this method, and for having taught it to me so that I can share it here with you.

PREPARATION

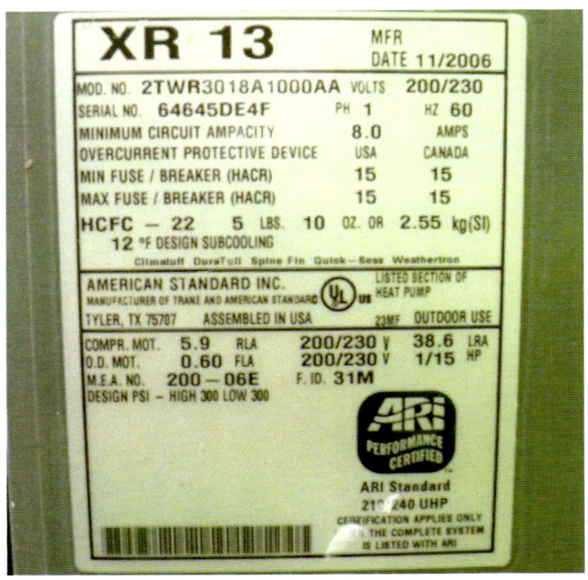

Determine which refrigerant the system uses by reading the manufacturer's label on the condenser unit, outside. You'll be using the corresponding data on the superheat table.

Record the manufacturer's design subcooling, also stated on the outdoor condenser unit (typically 10–12°F).

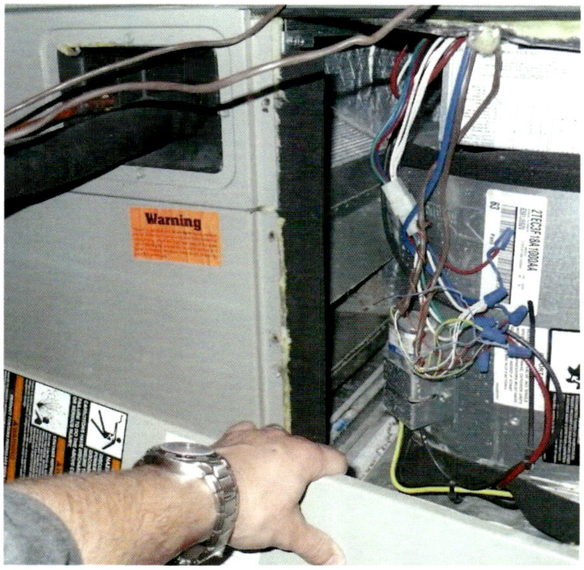

Determine whether the system's metering device is TXV or fixed by reading the manufacturer's labels on the equipment or inspecting the evaporator coil.

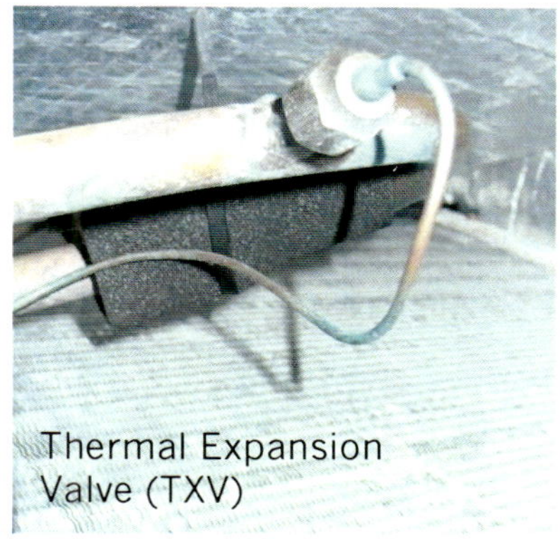

Finally, evaluate the efficiency the system, based on the characteristics in **Figure 2.13.2**, and note what design temperature difference you're expecting to see. ***TIP:** *It may take time to begin to recognize these characteristics, but practice makes perfect. Also, don't pay any attention to the words 'High Efficiency' on the labels; it's usually just B.S.*

Figure 2.13.2

	Standard Efficiency	**High Efficiency**	**Ultra High Efficiency**
R-22	*Standard size evaporator* *Standard size condenser* *Fixed orifice metering device*	*Standard size evaporator* *Bigger condenser* *TXV metering device*	*Bigger evaporator* *Bigger condenser* *TXV metering device*
R-410A	–	*Standard size evaporator* *Much bigger condenser* *TXV metering device*	*Bigger evaporator* *Much bigger condenser* *TXV metering device*

Standard Efficiency

Evaporator coil is designed to be 35°F colder than the return air

Condenser coil is designed to be 30°F warmer than outdoor air around it

High Efficiency

Evaporator coil is designed to be 35°F colder than the return air

Condenser coil is designed to be 25°F warmer than outdoor air around it

Ultra High Efficiency

Evaporator coil is designed to be 30°F colder than the return air

Condenser coil is designed to be 20°F warmer than outdoor air around it

Figure 2.13.3

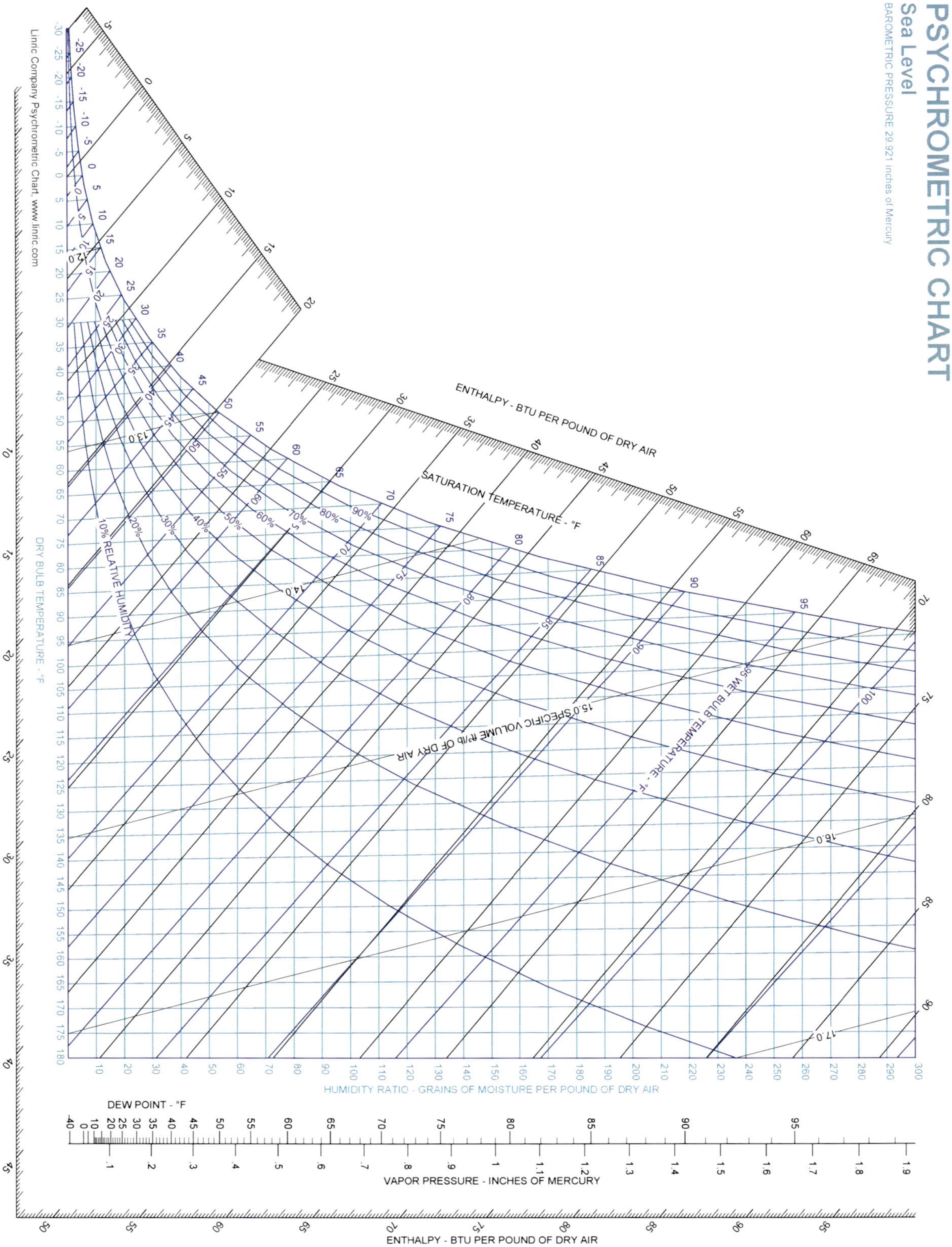

TESTING

Turn on the A/C (*see A.02*), and let it run for 10 minutes. Record the wet bulb and dry bulb temperatures in the return plenum. Alternatively, measure the dry bulb temperature and relative humidity in the return plenum and calculate the wet bulb temperature using **Figure 2.13.3** or the Psychrometric Calculator at:

www.BuildingPerformanceWorkshop.com/calculator

FOR TXV SYSTEMS: Measure and record the temperature of the suction line (the bigger, insulated pipe) as it comes out of the supply plenum, using a clamp thermometer or a contact thermocouple under the insulation. You have now measured the saturation temperature plus EVAPORATOR superheat.

FOR NON-TXV (FIXED) SYSTEMS: Measure and record the temperature of the suction line (the bigger, insulated pipe) at the outdoor suction service valve, using a clamp thermometer or a contact thermocouple under the insulation. You have now measured the saturation temperature plus TOTAL superheat.

***TIP:** *If there is no way around the pipe insulation, carefully cut the insulation in a straight line all the way around the pipe so it will still line up perfectly once you're done with the test. Slide the insulation to the side to attach the clamp thermometer, or insert the contact thermocouple.*

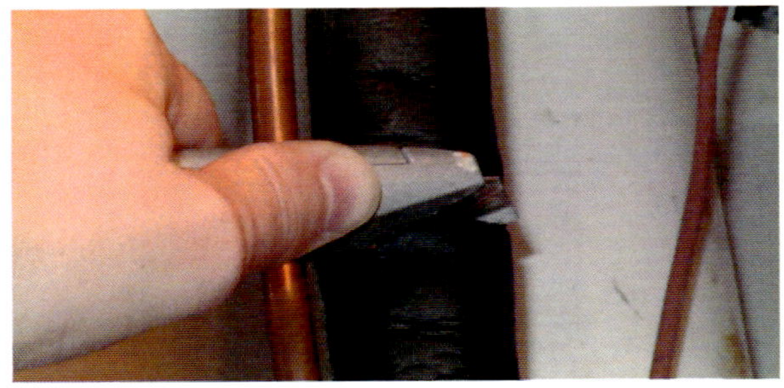

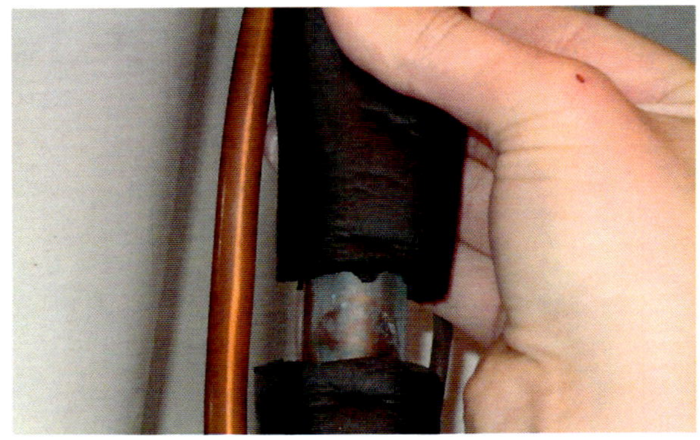

Measure the dry bulb temperature of the outdoor air going into the condenser unit (not coming out of it).

Measure and record the temperature of the liquid line (the smaller, uninsulated pipe) as it comes out from the condenser coil, using a clamp thermometer or a contact thermocouple.

INTERPRETATION

First, let's deal with the simpler side: subcooling of the liquid line. Calculate the design saturation temperature of the condenser coil by using **Equation 2.13.1**:

Equation 2.13.1 $$\boldsymbol{Temp_{Condenser} = DryBulb_{Outdoors} + \Delta T_{Design}}$$

Where:

$Temp_{Condenser}$ = Saturation temperature of the liquid line/condenser coil [°F]

$DryBulb_{Outdoors}$ = Temperature of air coming into condenser coil [°F]

ΔT_{Design} = Design temperature difference between the outdoor air and the condenser coil, from **Figure 2.13.2** [°F]

Next, use **Equation 2.13.2** to determine the temperature minus the subcooling, which is the temperature that should ideally be measured at the liquid line. You will then know what the liquid line temperature SHOULD be, and can compare that with what it actually IS. You want the measured subcooling effect to be within +/–3°F of the target if using digital tools, or within +/–5°F for analog tools.

Equation 2.13.2 $$\boldsymbol{Design_{Liquid} = Temp_{Condenser} - Subcool_{Target}}$$

Where:

$Design_{Liquid}$ = design temperature of the liquid line at present [°F]

$Temp_{Condenser}$ = Saturation temperature from **Equation 2.13.1** [°F]

$Subcooling_{Target}$ = Target subcooling, taken from the equipment label [°F]

Example:

$DryBulb_{Outdoors}$ = 84.9°F

ΔT_{Design} from **Figure 2.13.2** for our standard efficiency A/C = 30°F

$$\boldsymbol{Temp_{Condenser} = 84.9 + 30 = 114.9°F}$$

$Temp_{Condenser}$ = 114.9°F

Subcooling$_{Target}$, as stated on equipment label = 12°F

$$\boldsymbol{Design_{Liquid} = 114.9 - 12 = 102.9°F}$$

Measured liquid line temperature = 101.8°F (within +/–3°F, so we're good!)

Now for the superheat of the suction line; use **Equation 2.13.3** to calculate the design temperature of the evaporator coil:

Equation 2.13.3 $$\boldsymbol{Temp_{Evaporator} = DryBulb_{Indoors} - \Delta T_{Design}}$$

Where:

$Temp_{Evaporator}$ = Saturation temperature of the suction line/evaporator coil [°F]

$DryBulb_{Indoors}$ = Air temperature in the return plenum [°F]

ΔT_{Design} = Design temperature difference between return air and evaporator coil, from **Figure 2.13.2** [°F]

And finally, use **Equation 2.13.4** to calculate the design temperature of the suction line just past the evaporator coil:

Equation 2.13.4 $$\boldsymbol{Design_{Suction} = Temp_{Evaporator} + Superheat_{Target}}$$

Where:

$Design_{Suction}$ = design temperature of the suction line at present [°F]

$Temp_{Evaporator}$ = Saturation temperature of the suction line/evaporator coil [°F]

$Superheat_{Target}$ = Target superheat, taken from the table in **Figure 2.13.1** (for TXV systems, target evaporator superheat = 6–12°F)

Example:

$DryBulb_{Indoors}$ = 75°F

ΔT_{Design} from **Figure 2.13.2** for our standard efficiency A/C = 35°F

$$\boldsymbol{Temp_{Evaporator} = 75 - 35 = 40°F}$$

$Temp_{Evaporator}$ = 40°F

$Superheat_{Target}$ for our non-TXV system from **Figure 2.13.1** = 8.5°F

(as in the photos: 85°F outdoor dry bulb, 62°F indoor wet bulb)

$$\boldsymbol{Design_{Suction} = 40 + 8.5 = 48.5°F}$$

Measured suction line temperature = 51.3°F (within +/–3°F, so we're good!)

If either the liquid or suction line temperatures were not within +/–3°F of the design temperature using digital tools, or within +/–5°F for analog tools, then you have diagnosed a potential opportunity for improvement in the refrigerant system, and you should either hand the test results off to an HVAC specialist, or proceed to 2.13-B if you are an HVAC specialist.

2.13-B Pressure Gauge Method

Materials needed:

- A/C manifold pressure gauge set (low loss couplings recommended)
- wet bulb/dry bulb probe thermometer
- clamp or contact thermocouple (thermometer)
- safety glasses and gloves

CONCEPT

At a minimum, an EPA 608 Certification is required for professionals to run this test method, and if you successfully identified a refrigerant issue with **Method 2.13-A**, the pressure gauges won't tell you anything more unless you're an HVAC specialist, anyway. Always run Method 2.13-A before resorting to gauges, and ALWAYS make sure both the condenser and evaporator coils are perfectly clean, and verify the total system airflow is correct according to 2.01.

PREPARATION

Run 2.13-A, Non-Invasive Method to collect information about the system (TXV, non-TXV, efficiency and refrigerant type), and measure the return air wetbulb, outdoor air drybulb, liquid line, and suction line temperatures.

TESTING

At the outdoor condenser unit, locate the suction service valve (it's on the thicker of the two lines, and is generally insulated). Don your safety glasses and gloves, and unscrew the small nut covering the suction valve.

Ensure every knob and coupling on your gauge set is tightly closed. Connect the low-pressure hose (generally blue) from the manifold gauge set to the suction valve by screwing it firmly in place. Watch the corresponding gauge's reading of the pressure in the suction line.

Double-check that every valve on your gauge set is tightly closed. Connect the high-pressure hose (generally red) from the manifold gauge set to the liquid service valve (it's on the thinner, uninsulated of the two lines). Watch the corresponding gauge's reading of the pressure in the liquid line. The A/C has now been running for over 10 minutes, and should have reached a steady state.

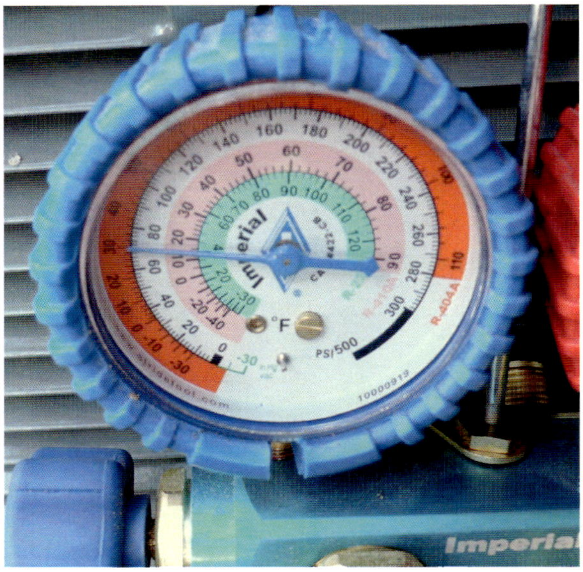

Once the low side gauge stops moving, record the suction line pressure and saturation temperature for the system's specific coolant on the gauge (in this case, 42°F for this R-22 system).

Once the high side gauge stops moving, record the liquid line pressure and saturation temperature for the system's specific coolant on the gauge (in this case, 88°F for this R-22 system).

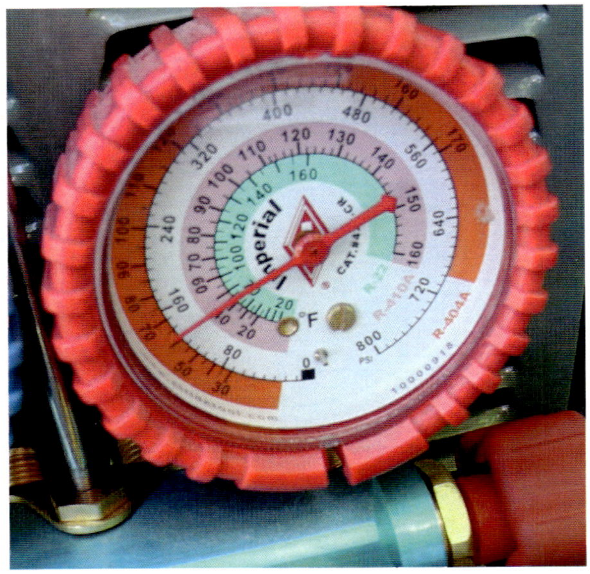

HEADS UP IF YOU USE THE HACK DISCONNECTION TECHNIQUE: If you disconnect a standard hose from the liquid service valve, do it quick, since the refrigerant will be spraying all over the place. This is totally avoidable by using low-loss couplings. **WARNING:** It's very important to be wearing the safety glasses and gloves at this point.

***TIP:** *If you have low-loss couplings, the refrigerant in the hose will be retained when you disconnect it from the service valve. If you have standard hoses, you can always get low-loss adapters.* Disconnect the high side hose from the liquid service valve; you now have a few ounces of the client's refrigerant in your hose.

To recover the refrigerant in the high side hose back into the A/C system, first open the low side valve on the gauge set to the central manifold while it's connected to the suction line. This pushes any air and contaminants that were present in the hose into the manifold. Quickly and carefully open and close the Refrigerant (REF) valve (connected to the thinner of the two hoses) to release a tiny hiss of air and contaminants from the manifold, and purify the tool. Then firmly re-tighten all the valves closed.

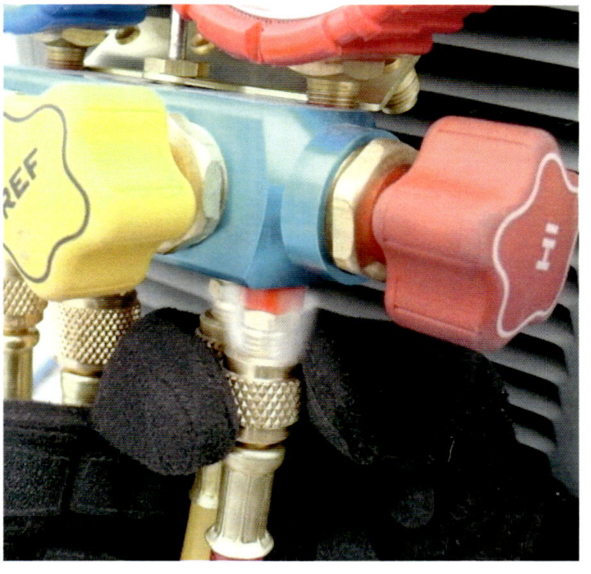

Carefully loosen the coupling between the disconnected high side hose and the gauge until a tiny amount of refrigerant hisses out; this pushes any air and contaminants out of the hose to avoid contamination of the refrigerant system.

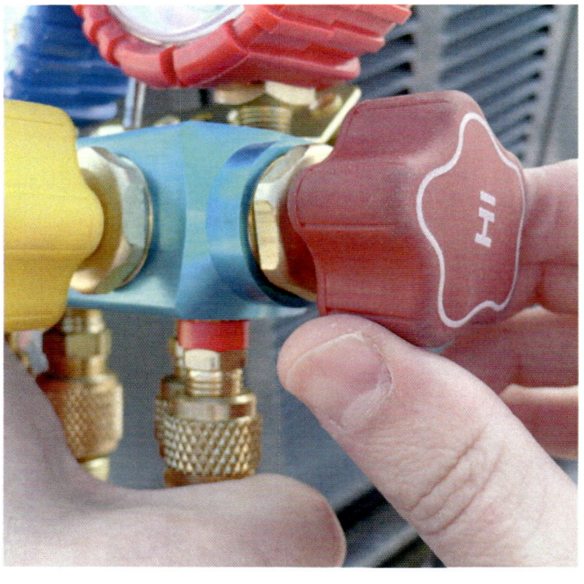

With all other valves tightly closed, open the disconnected high side hose's valve to push the liquid refrigerant from the hose into the central manifold.

Then open the low side valve to push the higher-pressure liquid refrigerant back into the lower-pressure suction line. Close all the valves again, and disconnect the low side hose from the suction service valve.

INTERPRETATION

For subcooling: Use **Equation 2.13.5** to calculate the actual subcooling:

Equation 2.13.5

$$\boldsymbol{Subcooling_{Actual} = Temp_{Saturation} - Temp_{Liquid}}$$

Where:

$Temp_{Saturation}$ = Saturation temperature of the liquid line [°F]

$Temp_{Liquid}$ = Measured temperature of the liquid line at the service valve[°F]

If the actual subcooling is more than +/–3°F off the target when using digital tools, or +/–5°F using analog tools, there is an opportunity for improvement in the refrigerant loop. If it's too low, it may be undercharged, and the refrigerant is condensing too late in the condenser coil; if it's too high, it may be overcharged, and the refrigerant is condensing too early in the coil.

Example:

$Temp_{Liquid}$ = 81°F measured at the compressor

$Temp_{Saturation}$ = 88°F

$$\boldsymbol{Subcooling_{Actual} = 88 - 81 = 7°F}$$

Design subcooling stated by this manufacturer was 12°F, and since we're within the +/–5°F accuracy range for our analog tools, we're good! However, since it's on the low side, we should keep an eye on this client's equipment as we continue to rock their world with our awesome diagnostic services.

For superheat: If the system you're testing is TXV-metered (which you discovered in 2.13-A), checking superheat is good, though it shouldn't change from the design 6–12°F for evaporator superheat (which is measured at the suction line at the evaporator coil, inside, according to 2.13-A). Total superheat (which includes the superheat from the evaporator AND the rest of the suction line, and is measured at the outdoor suction service valve) for a TXV system should be in the range of 12–20°F (again, with an accuracy of +/–3°F with digital tools, or +/–5°F with analog tools).

For non-TXV systems, calculate the actual superheat using **Equation 2.13.6**:

Equation 2.13.6

$$Superheat_{Actual} = Temp_{Suction} - Temp_{Saturation}$$

Where:

$Temp_{Suction}$ = Temperature of the suction line at the service valve [°F]

$Temp_{Saturation}$ = Saturation temperature of the suction line [°F]

The actual superheat can then be compared with the target total superheat taken from the equipment manufacturer's documentation, or the two-page table in **Figure 2.13.1**. If the actual superheat is more than +/–3°F off the target when using digital tools, or +/–5°F using analog tools, there is an opportunity for improvement in the refrigerant loop. If it's too low, it may be overcharged, and the refrigerant is evaporating too late in the evaporator coil. If it's too high, it may be undercharged, and the refrigerant is evaporating too early in the coil.

Example:

$Temp_{Suction} = 51°F$

$Temp_{Saturation} = 42°F$

$$\boldsymbol{Superheat_{Actual} = 51 - 42 = 9°F}$$

Target superheat for a 61°F day with a 57°F indoor wet bulb temperature is 16.3°F, and we are more than 5°F below this target with our analog gauges. This, plus the low reading on the subcooling from the previous example, might point to a dirty condenser coil, except for the fact that you cleaned that bad boy as soon as you got to the job, right?

Figure 2.13.4 a table of diagnostic results and their possible causes, developed by TruTech Tools:

System Problem (Fixed)	Suction Pressure	Liquid Pressure	Superheat	Subcooling	Amps	R/H Control
Refrigerant Overcharge	↑	↑	↓	↑	↑	Poor
Refrigerant Undercharge	↓	↓	↑	↓	↓	ICE/Poor
Liquid Restriction (Dryer)	↓	↓	↑	↑	↓	ICE/Poor
Low Evaporator Airflow (low load)	↓	↓	↓	↓	↓	ICE/Normal
Dirty Condenser	↑	↑	↓	↓	↑	Poor
Low Outside Ambient Temperature	↓	↓	↑	↑	↓	ICE/Normal
Inefficient Compressor$_3$	↑	↓	↑	↑	↓	Poor
Non-Condensables	↑	↑	↓	Normal$_6$	↑	Poor

System Problem (TXV)	Suction Pressure	Liquid Pressure	Superheat	Subcooling	Amps	R/H Control
Refrigerant Overcharge	Normal	↑	Normal	↑	↑	Normal
Refrigerant Undercharge	Normal$_2$/↓	↓	Normal$_2$/↑	↓	↓	ICE/Normal
Liquid Restriction (Dryer)	↓	↓	↑	↑	↓	ICE/Poor
Low Evaporator Airflow (low load)	↓	↓	Normal$_4$	Normal	↓	ICE/Normal
Dirty Condenser	Normal	↑	Normal	Normal	↑	Normal
Low Outside Ambient Temperature	Normal$_4$	↓	Normal$_4$	Normal	↓	Normal
Inefficient Compressor$_3$	↑	↓	↑	↑	↓	Poor
TXV Bulb Loose	↑	↑	↓	↓$_5$	↑	Poor
TXV Bulb Lost Charge	↓	↓	↑	↑	↓	ICE/Poor
TXV Bulb Poorly Insulated	↑	↑	↓	↓	↑	Poor
Non-Condensables	Normal	↑	Normal$_4$	Normal$_6$	↑	Poor/Normal

Figure 2.13.4

QUIZ Superheat and Subcooling

1) Why is it imperative that system airflow is verified and all issues addressed prior to running this test? What is this test trying to diagnose? ____________

____

____

2) Why should the non-invasive method always be used first? ________________

____

____

3) The measured dry bulb temperature in the return plenum is 69.1° F and the Relative Humidity is 39.8%. Using **Figure 2.13.3**, what is the wet bulb temperature? At what altitude is **Figure 2.13.3** designed for? Using the psychometric calculator from the Building performance workshop, how does altitude affect the wet bulb temperature? Is the change significant?

____

____

____

4) On a hot day, you measure the dry bulb temperature of the air entering the very large outdoor condensing unit and find it to be 96.9° F. Upon inspection of the indoor coil, you found a TXV and noted that the manufacturer's label indicated that the refrigerant is R-410A and the design subcooling is 10°F. You then measure the liquid line as it exits the condensing unit and record a temperature of 110.6° F. If you are using digital tools, does this system perform as you would expect? How about if you were using analog tools? In each case, would you assume there is a problem with the refrigerant? Assume the system has no airflow issues.

____

____

5) On a hot day, you measure the dry bulb temperature of the air entering standard sized outdoor condensing unit and find it to be 88.9° F. Upon inspection of the indoor coil, you do not find a TXV and noted that the manufacturer's label indicated that the refrigerant is R-22 and the design subcooling is 11°F. You then measure the liquid line as it exits the condensing unit and record a temperature of 100.6° F. If you are using digital tools, does this system perform as you would expect? How about if you were using analog tools? In each case, would you assume there is a problem with the refrigerant? Assume the system has no airflow issues.

6) On a hot day, you measure the dry bulb temperature of the air in the return plenum and find it to be 69.3° F. Upon inspection of the standard sized evaporator coil, you found a TXV and noted that the manufacturer's label indicated that the refrigerant is R-410A. You then measure the suction line as it exits the supply plenum and record a temperature of 49.8° F. Does the superheat of this type of system fall within acceptable parameters? Would you assume there is a problem with the refrigerant? Assume the system has no airflow issues.

7) On a hot day, you measure the dry bulb and wet bulb temperatures of the air in the return plenum and find them to be 73.9° F and 65.2°F, respectively. You then measure the outdoor dry bulb temperature entering the condenser and find it to be 93.1°F. Upon inspection of the standard sized evaporator coil, you did not find a TXV and noted that the manufacturer's label indicated that the refrigerant is R-22. You then measure the suction line at the outdoor suction service valve and record a temperature of 50.5° F. Does the superheat of this type of system fall within acceptable parameters? Would you assume there is a problem with the refrigerant? Assume the system has no airflow issues.

8) When using the pressure gauge method, describe the process of recovering the refrigerant in the 'high' side of the hose and putting it back into the A/C system.

9) What is happening with the refrigerant if the actual subcooling is too low? What about if it is too high? When is it considered to be too low or too high?

10) What is happening with the refrigerant if the actual superheat is too low? What about if it is too high? When is it considered to be too low or too high? ________________

____

____

____

____

11) You measure the liquid line at the service valve with a clamp thermometer and find it to be 96°F. With a set of pressure gauges, you discover that your R-22 system has a saturation temperature of 104°F. The manufacturers label indicates a design subcooling of 10°F. Is this system within acceptable limits? If not, what might be happening? ________________

____

____

____

____

12) For a non-TXV system, you measure the suction line at the service valve with a clamp thermometer and find it to be 56°F. With a set of pressure gauges, you discover that your R-22 system has a saturation temperature of 43°F. The condenser dry bulb temperature is 77°F and the return air wet bulb temperature is 58°F. Is this system within acceptable limits? If not, what might be happening?________________

____

____

____

____

Discussion and Critical Thinking:

1) Define superheat, subcooling, and saturation temperature. What role do these physical attributes of a refrigerant play in air conditioning systems? ______________________________

2) Research the refrigerants R-22 and R-410A. What are their similarities and differences? Are they both commonly used today? Why or why not?

____

3) What is a Thermal Expansion Valve? What purpose does it serve? How do you know if a system has one? Why are systems generally more efficient if they have one? ______________________________

4) What is an EPA 608 Certification? Are there different levels of certification? What is required to obtain one? ______________________________

5) Find a schematic of a complete A/C refrigerant cycle. Take note of which direction the refrigerant flows, locations of evaporator/condenser/compressor, typical temperatures in the lines, how airflow across these components is utilized, and other important features. Now, starting at the compressor, verbally discuss the journey

through the system that the refrigerant takes and how each component aids in reaching the end goal of conditioning the air. ________________

6) Take a look at **Figure 2.13.4** and discuss various system problems and their possible causes for fixed and TXV systems. Does is appear that superheat and subcooling diagnostics will help diagnose common problems? Why or why not? ________________

2.14 Gas Meter Clocking

Materials Needed:

- stopwatch
- calculator

PREPARATION

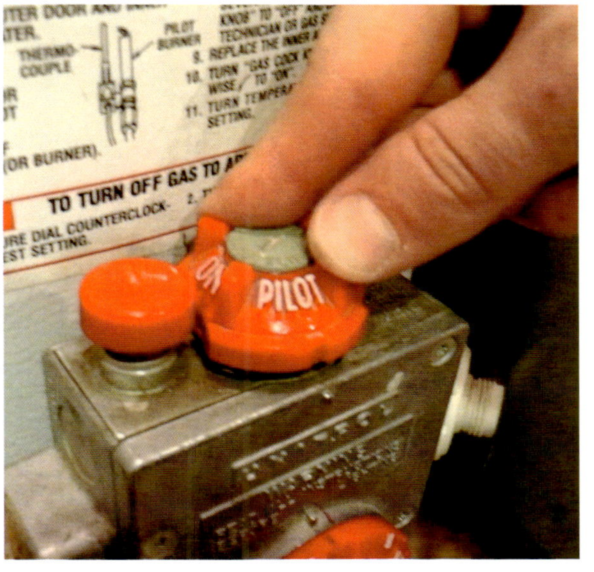

Turn off all gas fueled appliances, and locate the gas meter (may be indoors or outdoors).

Determine exactly how many Btu are contained per Cubic Foot of gas with the supplying utility company (for example, there are approximately 1,000 Btu per cubic foot of natural gas, though it will vary with supplier).

TESTING

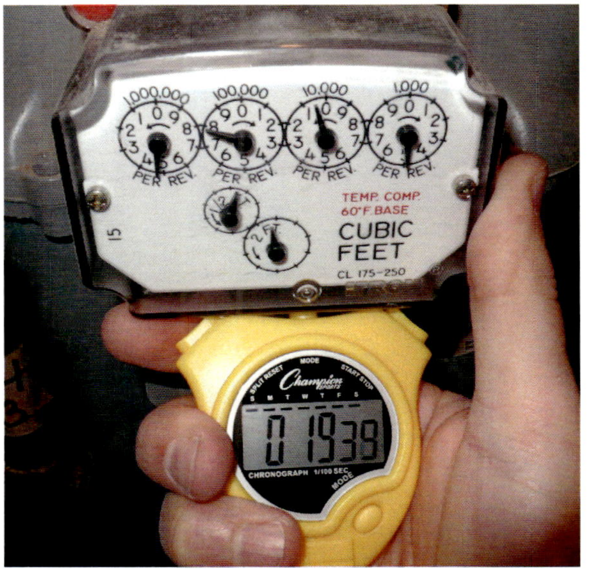

Verify that the gas meter is still, and not registering any gas use. Then fire on the combustion appliance you'd like to measure for gas usage, and return to the meter. Choose a dial to monitor for your time sample; you want a dial that isn't spinning like a madman (hard to keep track of), but you also don't want one that takes longer than 30 seconds to turn. Time a whole number of revolutions of the dial once the combustion appliance is running, for a minimum of 30 seconds. The longer you measure, the more accurate your numbers will be.

INTERPRETATION

Use **Equation 2.14.1** to calculate the amount of gas used per hour.

Equation 2.14.1 $$\boldsymbol{kBtu/hr} = \frac{\mathbf{3600} \times \boldsymbol{D} \times \boldsymbol{Local\ Btu} \times \boldsymbol{Rev}}{\mathbf{1000} \times \boldsymbol{t}}$$

Where:

D = Dial size [ft^3]

Local Btu = Btu content of utility's gas (if unknown, use 1000 for natural gas, or 2500 for propane) [Btu/ft^3]

Rev = Whole number of revolutions during the time sample

t = Time for one revolution of the meter [seconds]

Example:

t = 46 seconds

D = 1/2 ft^2 dial

Local Btu = 1,083 Btu/ft^3

Rev = 3 revolutions

$$\boldsymbol{kBtu/hr} = \frac{\mathbf{3600} \times .\mathbf{5} \times \mathbf{1083} \times \mathbf{3}}{\mathbf{1000} \times \mathbf{46}} = \mathbf{127.13\ kBtu/hour}$$

If your actual kBtu/hr measurement is within 10% of the stated input from the tested combustion appliance, everything's cool. If it's not, hand off the information to a qualified HVAC professional who can adjust the gas pressure or otherwise modify the system.

QUIZ Gas Meter Clocking

1) Why would you want a dial size that spins at a constant rate that is rotating at a medium speed? Why does accuracy increase with the duration of the measurement? ____________________

____

____

2) When clocking a gas combustion appliance, do you want to compare your results to the Btu input of the appliance, or the Btu output of the appliance? Why? ____________________

____

____

3) It takes 1 minute and 32 seconds for a 1 FT^3 dial to rotate 4 times after firing a 150-kBtu/hr furnace that burn natural gas. Given these parameters, does the consumption of the furnace cause alarm? The local Btu content is not known. ____________________

____

____

____

4) A portable natural gas furnace was converted to a propane furnace for ease of mobility, but there were no physical changes made to the burners. After using it few times, you discovered that it was burning at a very high temperature and decided to clock the gas meter. It takes 118 seconds for a ½ FT^3 dial to turn 3 complete revolutions. The local supply of propane contains 2516 Btu/FT^3. Given these parameters, does the consumption of the 75 kBtu/hr furnace cause alarm? ____________________

Discussion and Critical Thinking:

1) Why is it important to check with the local fuel supplier to determine the amount of Btu's contained in a cubic foot of gas? Does it vary significantly? How might this affect your calculation?

2) Discuss all of the components or parameters that affect how much gas can be consumed by a furnace or boiler. What can be adjusted or altered to reduce or increase the amount of gas (or Btu's) that is delivered to combustion site?

3) Look at the Btu content of several common fuels. Would it take more or less propane to heat a space than natural gas? How about methane? What do you notice between gaseous fuels and liquid fuels?

2.15 Gas Manifold Pressure

Materials Needed:

- gas pressure manometer (3-15inWC)
- hex key (3/32")

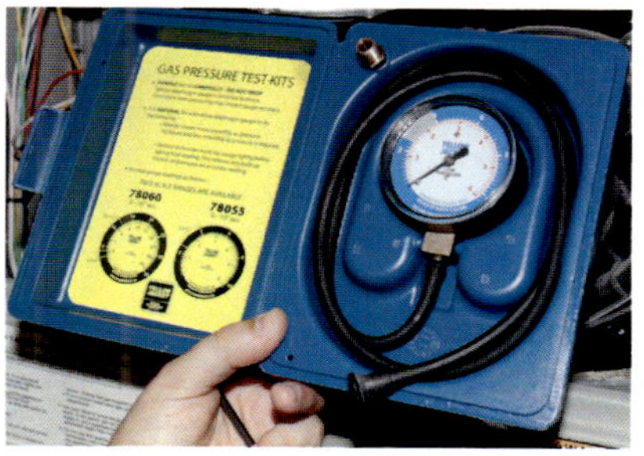

PREPARATION

Turn off the furnace/boiler at the service disconnect.

Turn off the gas valve (if possible) as an added safety precaution.

Verify the furnace/boiler's designed gas manifold pressure from the manufacturer's label (3.5inWC is a common design pressure for natural gas, and 10inWC is common for propane).

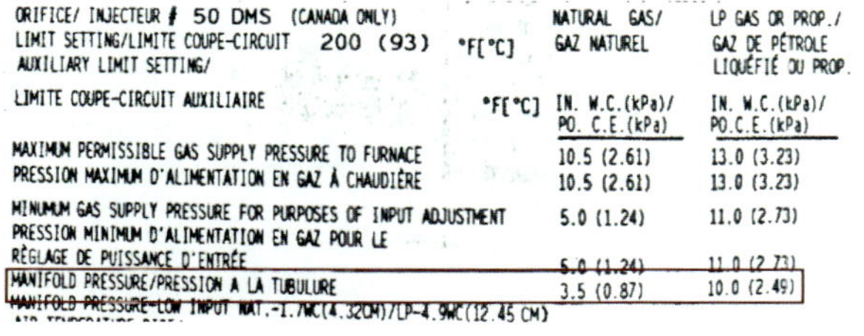

ORIFICE/ INJECTEUR # 50 DMS (CANADA ONLY)
LIMIT SETTING/LIMITE COUPE-CIRCUIT 200 (93) °F[°C]
AUXILIARY LIMIT SETTING/
LIMITE COUPE-CIRCUIT AUXILIAIRE °F[°C]

	NATURAL GAS/ GAZ NATUREL IN. W.C.(kPa)/ PO. C.E.(kPa)	LP GAS OR PROP./ GAZ DE PÉTROLE LIQUÉFIÉ OU PROP. IN. W.C.(kPa)/ PO.C.E.(kPa)
MAXIMUM PERMISSIBLE GAS SUPPLY PRESSURE TO FURNACE	10.5 (2.61)	13.0 (3.23)
PRESSION MAXIMUM D'ALIMENTATION EN GAZ À CHAUDIÈRE	10.5 (2.61)	13.0 (3.23)
MINUMUM GAS SUPPLY PRESSURE FOR PURPOSES OF INPUT ADJUSTMENT	5.0 (1.24)	11.0 (2.73)
PRESSION MINIMUM D'ALIMENTATION EN GAZ POUR LE RÈGLAGE DE PUISSANCE D'ENTRÉE	5.0 (1.24)	11.0 (2.73)
MANIFOLD PRESSURE/PRESSION A LA TUBULURE	3.5 (0.87)	10.0 (2.49)

MANIFOLD PRESSURE-LOW INPUT NAT.-1.7WC(4.32CM)/LP-4.9WC(12.45 CM)

Unscrew and set aside the bolt on the manifold side of the gas valve (closer to the burners).

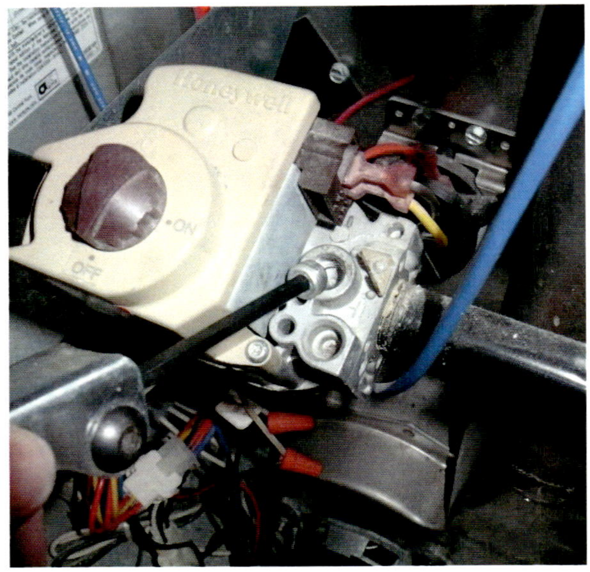

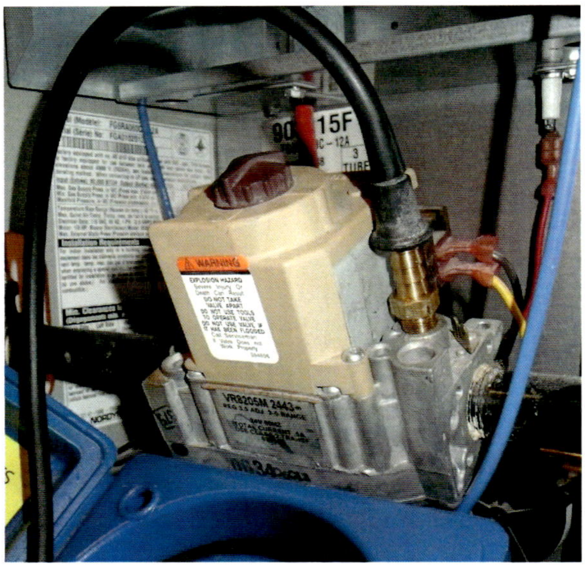

In its place, install the pressure tap; screw it down tightly using fingers and a wrench/pliers. Connect the end of the gas manometer's pressure hose to the pressure tap. **WARNING:** If the tap or hose connection is not tight, it will be a fiery disaster when you fire the furnace. Be very careful!

TESTING

Activate the furnace according to A.02 (call for heat and turn on the service disconnect switch).

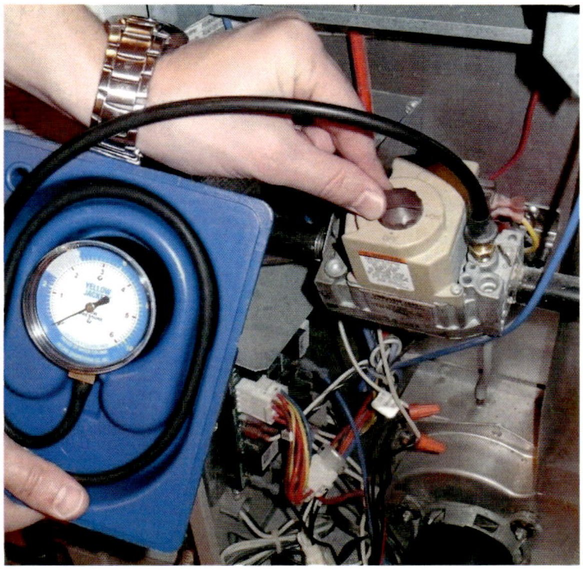

Turn on the gas valve. The furnace should still be running its startup sequence.

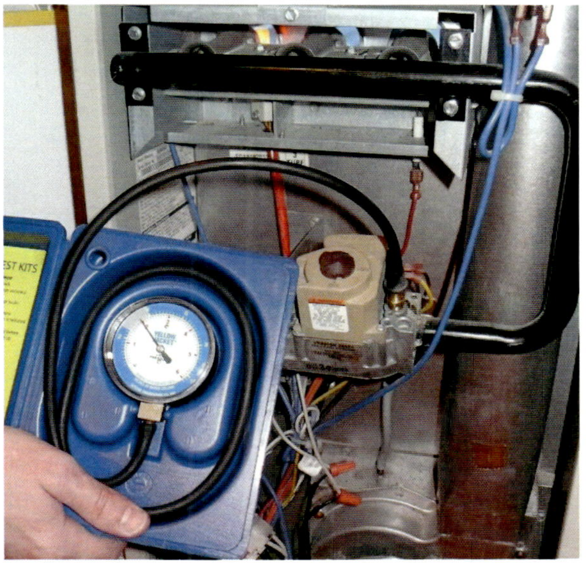

The gas valve will click open, and the pressure inside the manifold will show on the gauge. Compare this reading to the manufacturer's design pressure.

Turn off the furnace at the service disconnect, and turn off the gas valve as well, then proceed to disconnect the gauge and set the furnace up again for normal operation.

QUIZ Gas Manifold Pressure

1) Why is it a good idea to clock the meter for a furnace/boiler prior to testing the gas pressure at the manifold? ______________________________

____

____

2) Given the locality of the gas pressure manifold, why is it crucial to ensure that the tap and the hose are snugly fit? ______________________________

____

____

____

Discussion and Critical Thinking:

1) In regards to the manufacturers specified gas pressure, what would you assume is an acceptable deviation from this value? What if the actual pressure exceeds this deviation, yet after clocking the device, the furnace/boiler is consuming the correct amount of Btu's? ____________

____

____

2) Should the pressure equalize after the furnace reaches steady state? Would you be alarmed if the pressure fluctuated throughout testing?

____

____

3) What implications would a low gas pressure reading have for a furnace/boiler? What implications would a high gas pressure have?

____

____

3 ELECTRICAL DIAGNOSTICS

WARNING: Article 100 of the National Electric Code specifies that only 'Qualified People' should attempt to test electrical systems. Some of this stuff is dangerous.

3.01 Main Service Panel Circuit Testing

Materials Needed:

- rubber gloves and safety glasses
- electrical multimeter
- screwdriver
- utility knife
- infrared camera (optional)

PREPARATION

Open the door of the circuit breaker box or fuse box. Take note of any circuits that are in the 'OFF' position before touching anything; you can easily flip switches at this stage by accident. Don the safety gloves and glasses. Ensure the box cover is not painted stuck to the wall; use a utility knife to break the paint's seal if necessary.

Remove the box cover with your screwdriver: remove the bottom screws completely, and support the box cover with your other hand as you remove the top screws. Set the cover to the side. Double check the circuit breaker switches to ensure none were accidentally tripped during the removal of the cover. **WARNING:** You are in danger of being electrocuted at this point. Use caution.

TESTING

Scan for hot circuits with an infrared camera (or with the back of your bare hand, if you accidentally dropped your camera in the hot tub on your break). Pinpoint any hot circuits and identify their load location or appliance by their corresponding labels. ***HEADS UP:** *Some circuits and wires will be hot normally- what you're looking for is short spans or points of heat, rather than entire circuits or wires.*

Focus on the circuits/fuses you pinpointed earlier. Using the multimeter to detect Amps (current), attach the clamp around the single wire connecting to the circuit. ***TIP:** *You'll be detecting a current between 0 and 50 amps generally, so if your meter has multiple range settings, use a setting in this range. Residential circuits are all supplied with 120 or 240 Volts with 15 or 20 Amp circuit breakers or fuses.*

INTERPRETATION

If the circuit takes up one space, it should be running at approximately 120 Volts, and if it uses two spaces, it should be running at 240 Volts. Use **Equation 3.01.1** to calculate the electrical Power (in Watts) of the circuit:

Equation 3.01.1 $$W = E \times I$$

Where:

W = Power [Watts]

E = Voltage [Volts]

I = Current [Amps]

Example:

Your client has an indoor pool they suspect may be costing them tons of money. Let's test the pool heater to start with, before we get to the pump and the rest.

E = 120 Volts

I = 9.2 Amps

$$W = 120 \times 9.2 = 1104 \text{ Watts}$$

Use **Equation 3.01.2** to estimate energy use in Kilowatt-hours (kWh), the standard unit of electric consumption over time:

Equation 3.01.2 $$kWh = \frac{W}{1000} \times t$$

Where:

W = Power [Watts]

t = Time [hrs]

Example:

W = 1104 Watts

t = 6 hrs/day x 7 days/wk = 42 hrs/wk x 52 wks/year = 2184 hrs/year

$$kWh = \frac{1104}{1000} \times 2184 = 2,411 \text{ kWh per year}$$

If you know how much a Kilowatt-hour costs from your local utility, it's easy to use **Equation 3.01.3** to calculate exactly how much the estimated usage will cost:

Equation 3.01.3 $$\boldsymbol{Estimated\ Cost = kWh \times Price\ per\ kWh}$$

Example:

kWh = 1,722.24

Price = $0.15 per Kilowatt-hour

$$\boldsymbol{Estimated\ Cost = 2411 \times 0.15\ = \$361.67\ \text{per year}}$$

QUIZ Main Service Panel Circuit Testing

1) A client complains of high electricity bills during the winter and does not know why this is. Three individual circuits catch your attention so you use a multimeter to investigate further. During testing you find the following:

-Refrigerator:	5.5 Amps
-Electric Resistance Heater:	13.5 Amps
-Sump Pump:	8.8 Amps

Knowing this information, which of these presents the largest opportunity for improvement? Which of these devices would you expect to run the most in the winter? How about the summer? ______________________________

____

____

2) What is the total power that each device from question 1 is consuming?

____

____

____

3) Upon inspection, you find that the electric resistance heater is not controlled by a thermostat, but simply by an on/off switch. Your client indicates that the heater usually runs from December through end of March at the current setting. How many kilowatt-hours are consumed in this time frame? For this calculation, assume the heater has a constant draw of 1500 watts. Assume a non-leap year. ______________________________

____

____

4) You discover that the annual total consumption for the electric resistance heater is 4500 kWh. Local electricity is about $0.135 per kWh. How much is your client paying each year for this heater alone?____________________

____

____

3.02 Plug Load Wattages

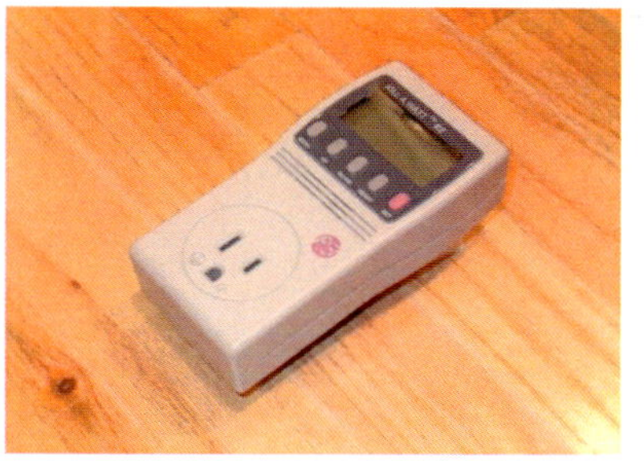

Materials Needed:

- power usage meter or datalogger

TESTING

Plug the electrical appliance to be tested into the power meter or datalogger, and configure the meter to calculate Watt-hours (Wh) or Kilowatt-hours (kWh), and estimated cost based on local electrical cost (if possible). Allow the meter to record as long a time sample as possible for greater accuracy. Refrigerators or other appliances that cycle on and off should be monitored for at least one full day of regular use.

QUIZ Plug Load Wattages

1) Why is it important to leave a plug load datalogger in place for long periods of time for appliances that have regular on/off cycles? ____________________

____

____

2) The nameplate of an appliance indicates it will draw 1100 Watts of power. Your datalogger, however, indicates 1020 Watts while operation. Does this surprise you? What are some factors that can affect this? ________________

____

____

Discussion and Critical Thinking:

1) A therm of natural gas has 100000 Btu's. How many kWh's of electricity will it take to produce the same amount of heat if 1 kWh contains 3413 Btu's per kWh? Lets assume that local cost dictate that gas is $1/Therm and that electricity is $0.10/kWh. Which is cheaper to create the same amount of heat? ____________________

____

____

2) For devices with variable controllers (i.e. lights, fans, motors, etc...), would you assume that power draw increases linearly with the speed of the device? Why or why not? ____________________

____

____

____

3.03 Electric Meter Clocking

Materials needed:

- stopwatch and calculator

TESTING

Turn off any major appliances that are not part of the baseline (for example, dishwashers, hot tubs, etc), and time a whole number of revolutions of the electric meter for a minimum of 30 seconds.

Alternatively, you can start the clock once you've recorded the exact positions of the five dials: these tell you the usage from left to right as a 5 digit number. For example, the meter in the photo shows a beginning usage of 45,015 Kilowatt-hours (kWh). Then walk away and come back later to stop the clock and record the new position of the dials.

INTERPRETATION

There will be a Kh value for each revolution on the face of the meter (for instance, '7.2 Watt-hours per revolution' on analog meters). Use **Equation 3.05.1** to calculate the kWh usage per day of the baseload:

Equation 3.05.1
$$kWh\ per\ Day = \frac{864 \times D \times Rev}{t \times 10}$$

Where:

D = Dial size, in Watt-hours per revolution [Watt-hours/revolution]

Rev = Whole number of revolutions of the dial

t = Time elapsed [seconds]

Example:

D = 7.2 Watt-hours per revolution

Rev = 3 revolutions

t = 55 seconds

$$kWh = \frac{864 \times 7.2 \times 3}{55 \times 10} = 33.93\ kWh\ per\ Day$$

Use **Equation 3.05.2** to calculate the cost of the baseload to the client:

Equation 3.05.2
$$\$_{Baseload} = kWh_{Daily} \times Cost$$

Where:

kWh_{Daily} = Daily electric consumption from **Equation 3.05.1** [kWh]

Cost = Local utility cost per Kilowatt-hour [$]

Example:

kWh_{Daily} = 33.93 kWh

Cost = $.08

$$\$_{Baseload} = \mathbf{33.93 \times .08 = \$2.71\ per\ day}$$

Multiply the daily cost by 30 to get approximate cost per month, and by 365 for approximate cost per year.

QUIZ Electric Meter Clocking

1) Why does accuracy increase with the duration of the measurement? What can have an effect on short-term readings? ______________________________

____

2) Your client took a month long (31 days) vacation that corresponded with their electrical billing cycle. Upon return, they found that their bill was $44.96 with local prices at $0.10 per kWh. Their analog meter displays Kh=7.2 Wh per revolution. Assuming that only base load appliances were running while your client was gone and no customer fees are included, how long should it take for one revolution of the dial? ______________________

____

____

____

3) You have established that your client's electrical base load is about 9.9 kWh per day. What is their daily cost if local electrical prices are $0.16 per kWh? What is their estimated monthly and annual base load cost? _________

____

____

Discussion and Critical Thinking:

1) Which method do you feel would give a more accurate representation of the electrical base load: timing the rotating meter or reading the consumption dials at two intervals and clocking the times between the two readings? ____

____

2) List some common household items that are typically included in the electrical base load. ____

____

____

A APPENDIX: PREREQUISITES

A.01 Defining the Envelope

This may be the single most important task you will perform in any home. Defining the Envelope is a profoundly underrated step, and only the Whole Home Performance Professional can be trusted to do the job right.

A Quick Review:

The 'Envelope' is the boundary of conditioned space, and is made up of two components that must be perfectly aligned and uninterrupted:

- Pressure Boundary (or Air Barrier): the home's air sealing layer
- Thermal Boundary (or Thermal Barrier): the home's insulation layer

The envelope may not really be where it appears to be, however, because of:

- Air Bypasses (interruptions in the Pressure Boundary)
- Thermal Bypasses (interruptions in the Thermal Boundary)
- Separation of the Pressure and Thermal Boundaries
- misleading use of the space by the occupant (like removing the door to the unconditioned rear porch, or keeping the lawnmower in the basement)

And lastly, there are only 3 types of space:

- Conditioned Space
- Outdoors
- Buffer Zones (just like outdoors, except no precipitation, wind, or animals)

The Hard Part:

The builder, the homeowner, and the various tradespeople involved in a home all have an opinion about where the home begins and ends, and they are all generally wrong.

A ventilated attic or crawlspace may in fact be more connected to the home via air bypasses than it is to outdoors, and a 'finished' attic may look like it's inside, but may clearly prove to be a buffer space instead.

It's the first and most critical job of the home performance professional to define the envelope before you start recommending installation of air sealing or insulation in the wrong places. Use the Zonal Pressure Test (*see 1.03-A*) for this purpose.

A.02 HVAC Documentation and Controls

A.02-A Documentation

A.02-B Controls

A.02-C Tricking the HVAC into Cooling Mode in Cold Weather

A.02-A Documentation

On air conditioners, furnaces, boilers, and water heaters, you will generally find Btu capacity (input and/or output), efficiency, and date of manufacture. If not stated outright by the manufacturer, you can typically find the year of manufacture in the first four digits of the serial number (for example, 0201 would be either February 2001 or 2002 January).

The information below is a cursory overview, and ultimately there is no substitute for experience, which is your only tool in cases where there seems to be no pertinent information on the label, or contradictory information, or if the label is missing.

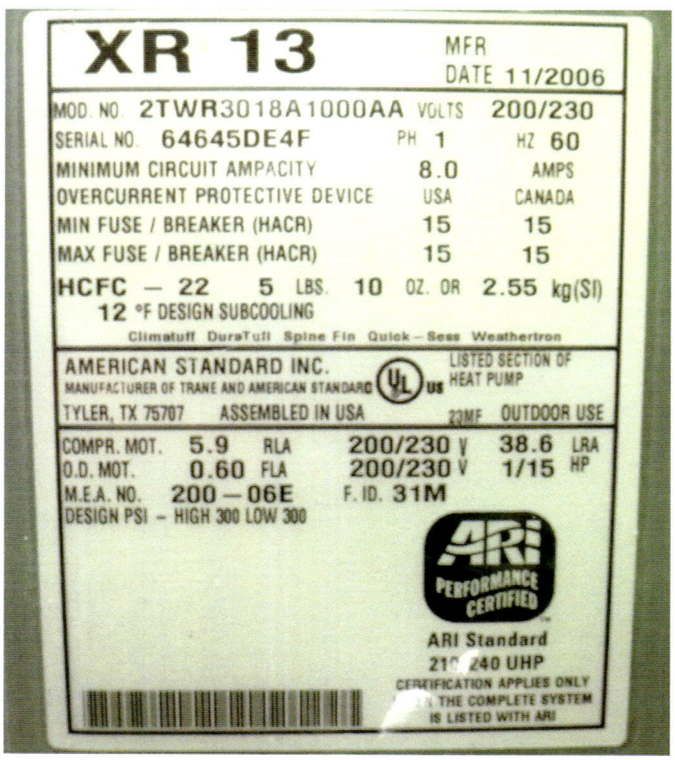

Figure A.02.1

Air conditioners and **Heat Pumps** for homes generally have detailed documentation on the condenser unit (outdoors) in a split system.

They are measured in tonnage; if the manufacturer doesn't state the tonnage in the model number, as they seem to have in **Figure A.02.1** (2 Ton), look for a number divisible by 12 or 6. There are 12,000 Btus per ton, so a model number containing an 18 would be a 1.5 Ton unit, and containing a 30 would be a 2.5 Ton unit.

The efficiency (EER, SEER, or COP) may be stated, as above (13 SEER), or may have to be ascertained separately. In 2005, the EPAct mandated a minimum SEER of 13, so if you know the age, you can deduce a ballpark estimate in the field. If the unit is high efficiency, they will brag to you the exact EER/SEER/COP, though the words 'High Efficiency' printed on the equipment are almost always meaningless advertising.

Furnaces and **boilers** are typically labeled with detailed information outside or inside their cabinets, as seen in **Figure A.02.2**. The Btu input and/or output may be buried in a dense paragraph of other information, but you will find one or the other. If only the input is listed, the output is calculated by multiplying the input by the efficiency (Annual Fuel Utilization Efficiency, or AFUE, which can be found on the yellow Energy Guide sticker). For example, a 60,000 Btu/h furnace that's 80 AFUE (or 80% efficient) would provide 48,000 Btu/h to the home.

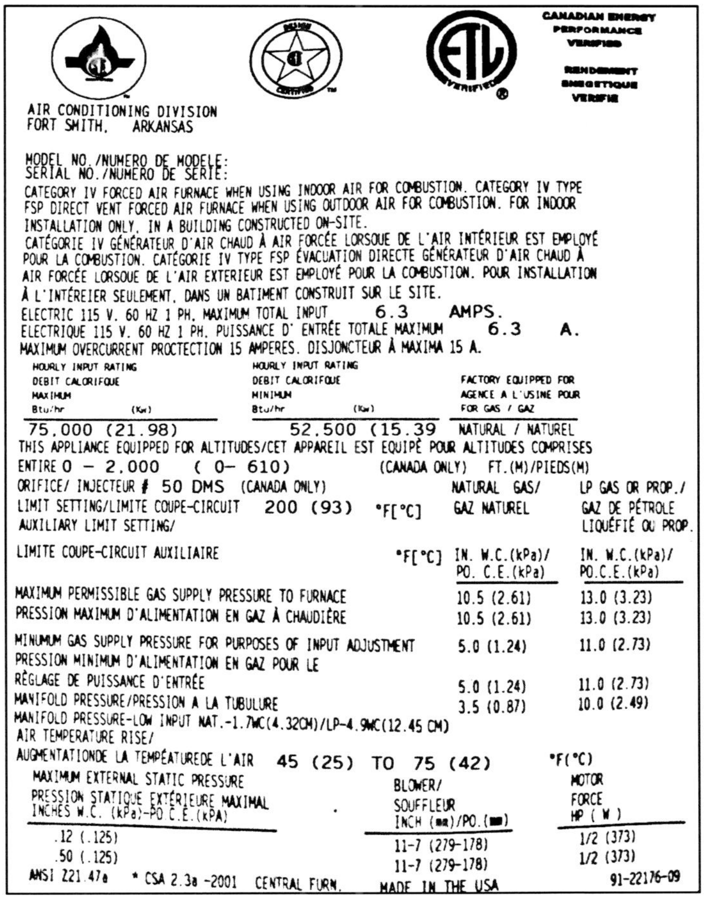

Figure A.02.2

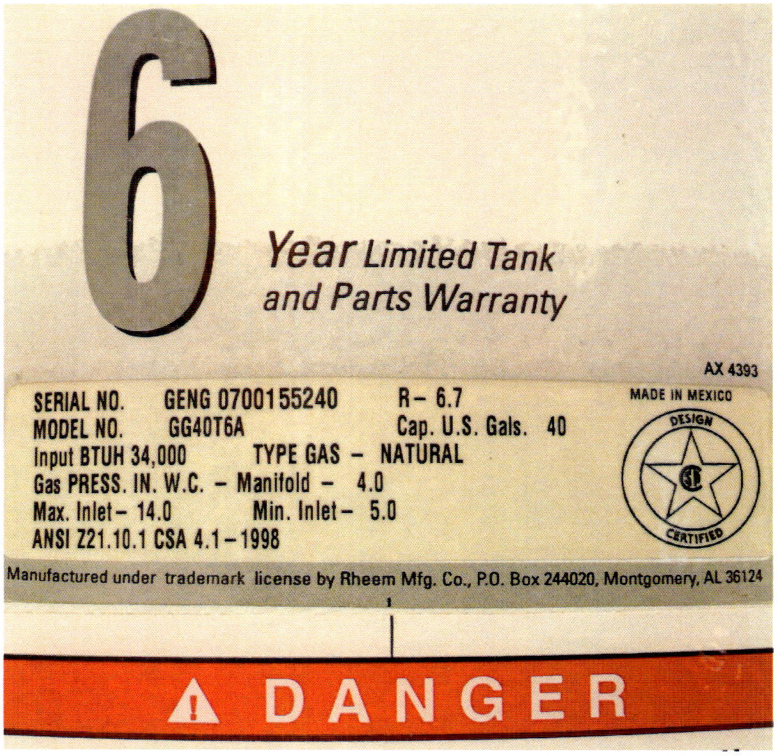

Figure A.02.3

Water heaters are labeled by size in capacity (not the same as the '1st Hour Rating' capacity), Btu/h input, and R-value of the tank insulation. As you can see in **Figure A.02.3**, the date of manufacture is usually in the first four numbers in the serial, and can be checked against the ANSI standard at the bottom of the label (the ANSI year will be the earliest year it could have been manufactured).

A.02-B Controls

Sometimes getting the air handler, furnace, or air conditioner to start or stop running can be tricky. Let's briefly review the controls for HVAC systems so you can confidently make the equipment do your bidding.

Service Disconnect (Emergency Kill Switch)

If you want to be sure nothing in the HVAC system will run, the service disconnect is what you should use. When this switch is off, nothing in the system (including the air handler or the water pumps) can run. It usually looks like a simple lightswitch, either mounted on the furnace/boiler or at the end of the electrical conduit coming out of the unit (you may have to hunt for it). It's possible a system does not have a master electrical switch of this type; in that case, you must use the thermostat to control the system.

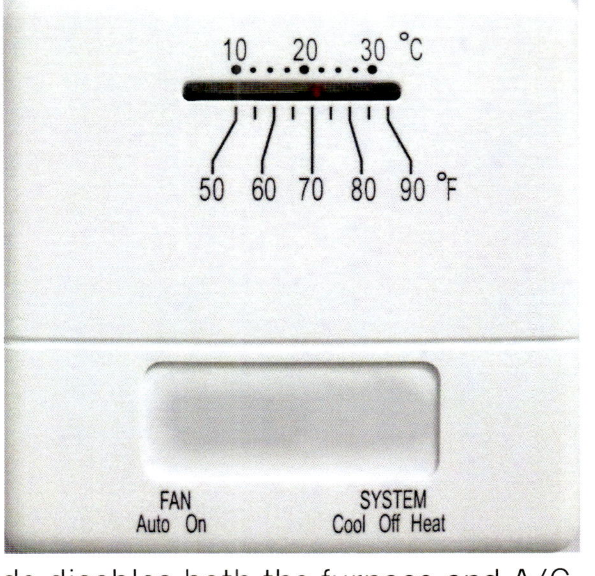

Manual Thermostat

Whether digital or analog, all manual (non-programmable) thermostats have three to four main mechanisms:

- SYSTEM (Heating/Off/Cooling)
- FAN (Auto/On)
- TARGET TEMPERATURE (55°–90°)
- MEASURED TEMPERATURE

The SYSTEM switch makes active the furnace in 'Heating' mode, and the air conditioner in 'Cooling' mode. The 'Off' mode disables both the furnace and A/C, but allows the air handler fan (or 'blower') to function. If your goal is to run the blower only, choose the 'Off' mode to avoid accidental heating or cooling.

The FAN switch can be set to 'Auto' mode for normal operation; the blower will only run when the thermostat calls either for heating or cooling. The 'On' mode will run the blower continuously; this mode is the only way to activate the blower if the SYSTEM switch is in 'Off' mode.

The TARGET TEMPERATURE slider sets the desired temperature for the home. Don't neglect to notice the temperature the homeowner has chosen before you adjust it; one useful technique is to adjust the target temperature in increments of 10° to easily remember the original target.

The MEASURED TEMPERATURE gauge shows the temperature around the thermostat. This can be an easy way to verify either the correct temperature in the home; or, if the reading is significantly different than what it should be, a thermostat that's located in a poorly-chosen spot (e.g. on a wall that holds a supply duct or has no top plate, etc).

Programmable Thermostat

Digital programmable thermostats have all the features of manual thermostats, and then some. They come in different shapes and complexities, but all of them share two common buttons:

- HOLD
- RUN

If you'd like to temporarily override the programming in order to run diagnostics, simply set the desired TARGET TEMPERATURE and press the HOLD button.

When you're finished with your testing and are ready to return the system to the homeowner's program, simply press the RUN button.

Water Heater Controls

Electric water heaters are simple to operate, and the only operational control is a thermostat, which is also a feature of gas fueled water heaters. Before adjusting the thermostat, make note of the occupant's chosen temperature (use a pen if it helps). Generally, the manufacturer's most clearly marked temperature setting is 120°F.

To fire a non-electric water heater by calling for heat, you can either turn up the thermostat or run hot water (a bathtub will be fastest).

The dial on top has three settings: 'On', 'Pilot', and 'Off'. You should never turn the dial to 'Off'; the 'Pilot' setting will prevent the unit from firing if you don't want it 'On'.

Newer water heaters may not have an obvious on/off switch or thermostat. If the power vent on top has a switch, use that; otherwise, unplugging the power vent will prevent the unit from firing. The thermostat may be hidden inside the body of the tank: look for a small panel with one or two screws, with a small slotted peg behind it, which can be turned with a small screwdriver to adjust the temperature.

Safety Controls

The only safety control we'll get into here is the pressure-temperature relief valve on a water heater or boiler. This valve is the primary safety device on the equipment, and as such it should be in working order. There are those who point this out, and then turn right around and say, 'Don't ever touch this, in case it breaks.' That's crazy talk, and you shouldn't listen to it. In fact, it states explicitly in most water heater manuals that the homeowner should open the PT valve regularly! At Chicago Home Performance, we test every single PT valve we see, and have never once been struck by lightning.

Gently pull the valve's lever until you hear a 'SSS', and immediately let go. Hopefully, the hissing sound of water escaping through the valve stops too, and a small amount of brown water spills out onto the floor.

If the hissing sound persists, or if the brown puddle on the floor continues to grow, the valve should be replaced, and it can temporarily be tapped gently back into place. When the valve sticks on its way back in, it may stick on its way out, and then it's no kind of safety device at all!

A.02-C Tricking the HVAC into Cooling Mode in Cold Weather

All air handlers have at least 3 speeds: High, Mid, and Low. The furnace and A/C use different speeds. When running total system airflow testing (*see 2.01*), using cooling mode is necessary, since the airflow for air conditioner systems is standardized (this has to do with physics and thermodynamics).

If the temperature outside is under 60°F, running the A/C compressor can damage the system, so you must trick the air handler into running in cooling mode, while not actually running the A/C compressor. There are a few different ways to do this, and here we'll demonstrate one of them:

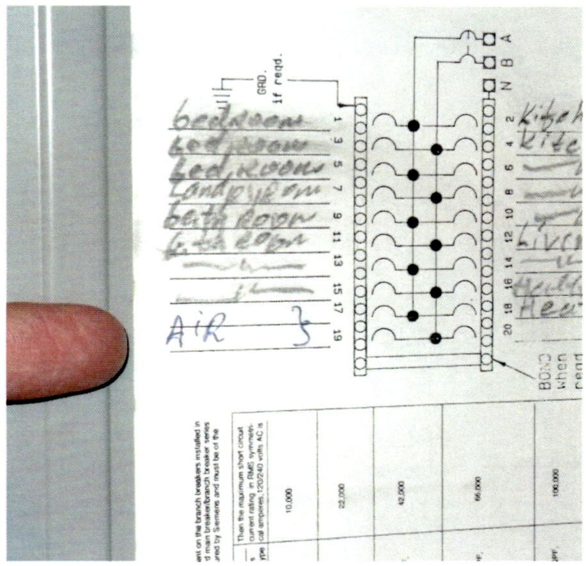

At the main electrical service panel, find the A/C compressor circuit. Make sure you can also see a separate 'Heat' or 'Furnace' label, so you know you've chosen the correct circuit, and deactivate it.

You may alternatively deactivate the compressor by pulling the outdoor service disconnect.

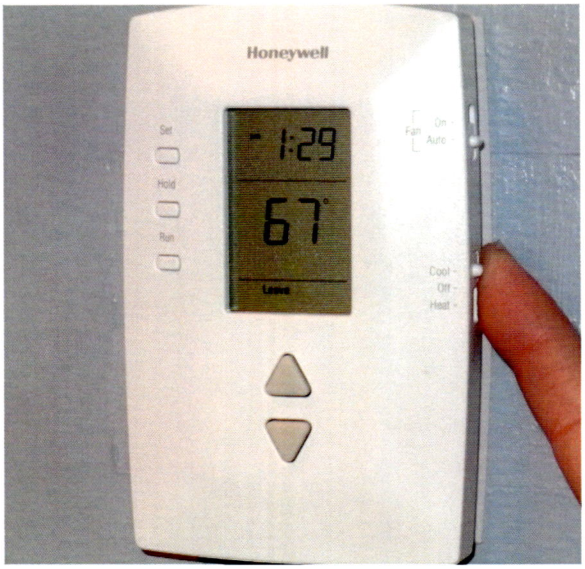

Turn the SYSTEM switch to 'Cool', and the FAN switch to 'On'.

Set the target temperature to below the measured home temperature so that the thermostat calls for cooling from the HVAC system, and if the thermostat is programmable, press the 'Hold' button. The air handler should now be running on cooling speed (if it's not, check to ensure the service disconnect switch is not turned off).

A.03 Manometer Mechanics

The digital dual-channel micromanometer is one of the most powerful tools in the arsenal of the home performance pro, but only if its mechanics are understood by the user. I've found that a huge number of professionals have not been properly instructed on the behind-the-scenes workings of this machine, so here's a start to correcting that.

Siamese Twins

The dual channel manometer is actually two devices slaved together into one case. If you think about them as Siamese twins, there's a smart twin and a dumb twin:

- Channel A (dumb twin) measures **pressure** only
- Channel B (smart twin) measures **pressure** plus **airflow**, **velocity**, and approximate cumulative **sizes** of air leaks

Both channels have an Input tap, Reference tap, and Display.

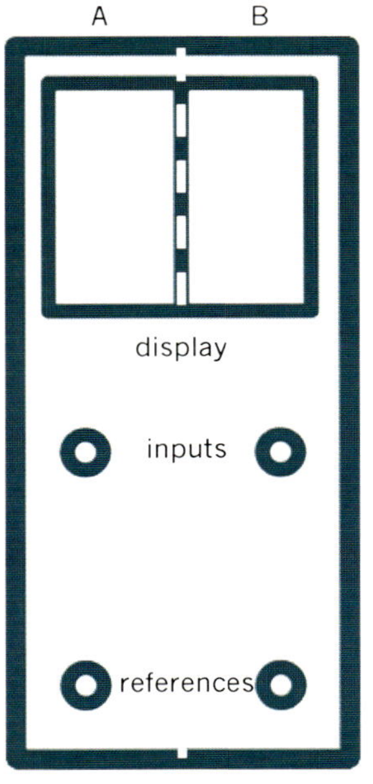

Say It Out Loud

There's a saying I teach all my students to use in the field:

"I'm measuring the pressure IN ________________

WITH REFERENCE TO ___________________."

Since you have an input tap and a reference tap, if you can clearly state what you want to measure, you can teach yourself to hook up the pressure hoses correctly.

Pressure is one thing exerting force on another thing. You can't just drop a hose on the ground and measure the pressure in the room around you. You need a barrier between one thing and another thing, for them to fight it out.

The Input tap gets exposed to whatever you want to measure the pressure IN, and the Reference tap gets exposed to whatever you want to REFERENCE.

Measuring Pressure

Pressure can be measured in Pascals (Pa) or Inches of Water Column (inWC). One Pascal can be thought of as the weight of a sticky note, and it takes 250 Pa to make 1inWC. Simply use the trick explained above to wire the manometer's hoses correctly, and you're halfway ready to measure.

A total pressure tap has only one opening; the taps on the manometer itself are total pressure taps, and so are the hoses once you connect them to the manometer. Airflow across this one opening affects the measurement greatly; therefore, make sure not to breathe on the manometer's taps when holding it in front of you, and protect the end of your pressure hose from airflow by placing it out of the way of fans, duct registers, door undercuts, etc.

Make sure your hoses have tight-fitting ends by cutting them off whenever they get stretched out from use. Also, inspect the hoses for punctures or other defects (we do this regularly as part of cleaning ours).

Measuring Airflow

As shown above, only Channel B measures flow, and it needs a separate device (a fan or flow meter) to do so. In the case of all of these devices, the manometer extrapolates airflow mathematically through the pressure difference before and after an opening. The opening is precisely calibrated for this purpose, and has an adjustable size; typically this size is called the 'Configuration' of the device (Open, Low Flow, Ring A/B/C, etc).

The bigger the opening, the more airflow is possible, at a lower velocity. Conversely, making the opening smaller allows less airflow but increases the flow's velocity.

The one last bit of important knowledge is that in flow mode, the Input tap is intended for the post-opening pressure (this is typically the a tap on the fan housing), and the Reference tap is intended for the pre-opening pressure.

The Name of the Test Reveals the Input

As a rule, the primary focus of any test will go on the Input tap. Many pros have a hard time remembering how to wire a manometer for a duct leakage test (*see 2.09 and 2.10*), but if you follow the rules in this section, you'll see that:

- The test is a measure of Pressure and Airflow
- Channel A only measures pressure, therefore Channel B must measure the flow
- You measure the pressure IN the ductwork with reference to (WRT) the home (evidenced by the name of the test)
- You measure the flow IN the fan with the Reference tap exposed to the same air as the intake of the fan

A.04 Drilling/Patching Holes

A.04-A Drilling in Single-Wall Combustion Flues or Sheet Metal Ductwork

A.04-B Drilling in Double-Wall Combustion Flues

A.04-C Drilling in PVC Combustion Flues

A.04-A Drilling In Single-Wall Combustion Flues (or Sheet Metal Ductwork)

Use one hand to solidly support the back side of the flue or duct you're drilling (they should not be hot at this point, for obvious reasons), and carefully apply pressure to the drill. Keep your hand off-axis from the drill's direction, in case you slip or the drill bit punches quickly through the flue. ***TIP:** *Use a unibit both for versatility and for safety: you won't pop the drill through the flue and injure your supporting hand, since the bit stops itself again and again. When using the unibit, if you're having a hard time getting the hole started, swivel the drill in a circular motion while drilling, to catch more of the edge of the bit.*

Combustion gases should always be sampled before dilution air is introduced, which means drilling a hole that will get your probe down into the throat of a water heater. Support the flue with a firm hand to avoid pushing it off-center. If the draft diverter is high enough for your probe to point down into the throat, or if your combustion analyzer's probe is flexible, no hole is necessary. The most convenient way to seal these holes is with high-temperature UL-181 aluminum tape, although you can also use sheet metal plugs or ASTM E-134 fireblock caulk (not ASTM E-814 firestop caulk or spray foam).

A.04-B Drilling in Double-Wall Combustion Flues

You can identify a double wall 'Type B' or 'Type L' vent by its crimped collar connector, the presence of writing or stickers (which would have burned off were it single-wall), the rivet connectors, and the absence of screws. The inner wall is stainless steel, and the outer wall is either galvanized or stainless steel; this wall insulates the inner one, giving a higher performance for draft.

While BPI national standards do not currently incorporate drilling into double-walled combustion flues, it can be done, and I'll demonstrate one way to successfully access the inside of the flue and re-establish the integrity of the inner and outer walls.

Fitting the drill with a large size unibit (1/4" to 3/4") will allow you to drill a large hole in the outer wall, and a small hole in the inner wall for your probe. The large outer wall hole gives better access for patching the inner hole. First, wrap your hand around the back of the flue to provide support. Then place the drill straight against the flue in an area near the collar, but clear of the lengthwise seam. Remember to always keep the drill off-axis from your hand to avoid injury.

Drill at medium speed until the unibit pops through to its first step-up. You have now made a 1/4" hole in the outer wall, and have not yet reached the inner wall.

You should clearly be able to see the inner wall through your initial hole.

Re-insert the drill bit in the hole, and drill on slow speed, pressing steadily into the flue while supporting it from the back with your hand. Each time the drill pops through another step-up to a bigger diameter, stop drilling and see if the inner wall has been punctured, until it is.

To patch the hole, first seal the inner wall's hole with ASTM E-136 fire block caulk (not ASTM E-814 fire barrier caulk or spray foam), and the outer wall with UL-181 high temperature aluminum tape.

A.04-C Drilling in PVC Combustion Flues

PVC flues are installed on combustion appliances that incorporate power venting or sealed combustion. There should always be two PVC pipes running from the furnace to outdoors: one brings in fresh air for combustion, the other exhausts the combustion gases to outdoors. It's critical to drill the PVC that's connected to the power vent fan, and not the one that's supplying fresh air from outdoors.

Support the back of the pipe with your hand, and rest the 1/4" drill bit against it, off-axis from your hand, to avoid accidental injury.

Using low speed, start drilling straight into the pipe until the bit has broken through the skin of the pipe and a small amount of shavings has built up.

Now tilt the drill downwards: a downward slope to the hole will ensure that condensed moisture inside the pipe will not flow outside the pipe. Continue to drill on low speed until the drill bit is forcibly pulled into the hole, and stop.

With the drill embedded in the wall of the PVC, reverse the drill's direction, as if you were loosening a screw. Drill on medium to high speed, and the drill bit pushes itself out of the hole. Pressing into the hole at the same time allows you to bore the hole to smoothness without danger of popping through the opposite side of the pipe.

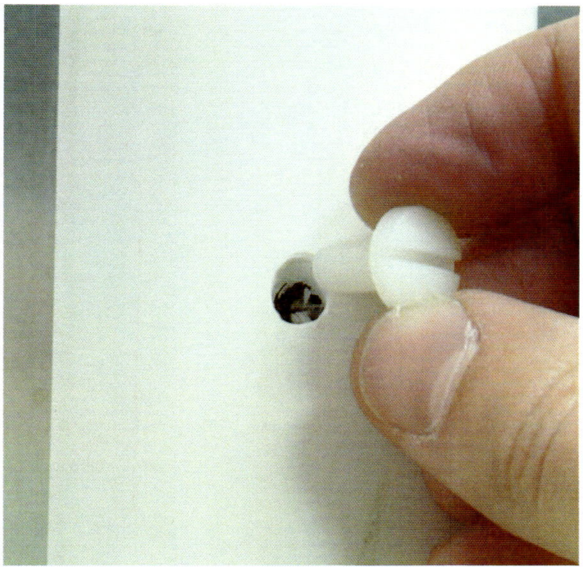

To plug the hole, use a 1/4" nylon or teflon screw.

A.05 'Baseline' Conditions

A home completely at rest, with minimal pressure differentials within its envelope, is said to be at 'baseline conditions'. Setting up baseline conditions is necessary for any accurate pressure-sensitive testing in the home. There are 3 main driving forces of pressure in a home:

- Stack Effect (the relationship between inside and outside)
- HVAC (air movement inside the home)
- Wind (air movement outside the home)

Baselining means separating inside and outside by:

- Closing the fireplace's chimney damper (the fire must be out first)
- Closing and locking doors and windows to the outdoors
- Opening all interior doors

  ***TIP:** *If you're unsure of which doors are interior and which are exterior (for example, side attics, three-season porches, rear stairwells), try to approximate wintertime conditions by asking the homeowner whether a door is generally closed or open through the winter months. See A.01 for more on 'Defining the Envelope'.*

Baselining also means eliminating the interior airflows and creating safe conditions for pressure testing by:

- Turning off the central HVAC system and all combustion appliances
- Turning off any exhaust fans and any other fans between indoors and outdoors

Once all the steps above have been taken, the home is in baseline conditions.

A.06 Dimensions and Geometry

Accurate spatial measurement of homes is critical for envelope diagnostics, ventilation sizing, pricing, and so on. You may consider using a laser distance meter to get the job done faster. Here are three techniques that can turn you into a home measuring machine:

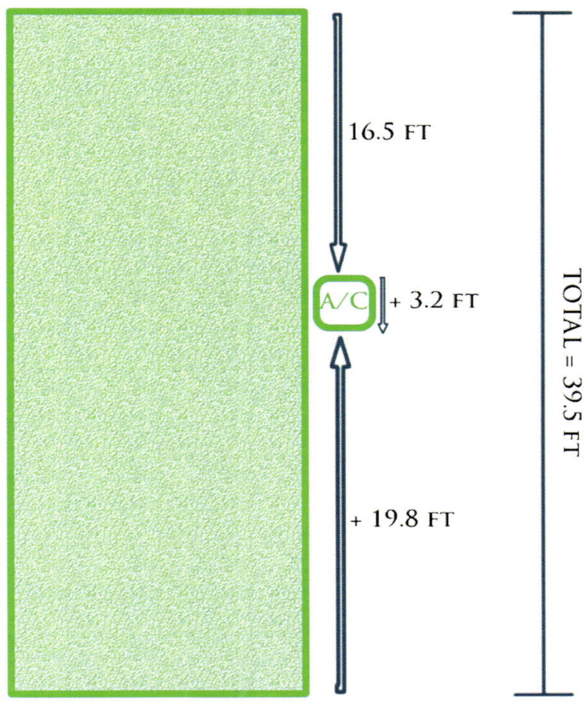

A.06-A Additive Method

When measuring a long distance (especially on a sunny day when the laser is difficult to see), use any features that interrupt the entire span, and add up the series of short measurements.

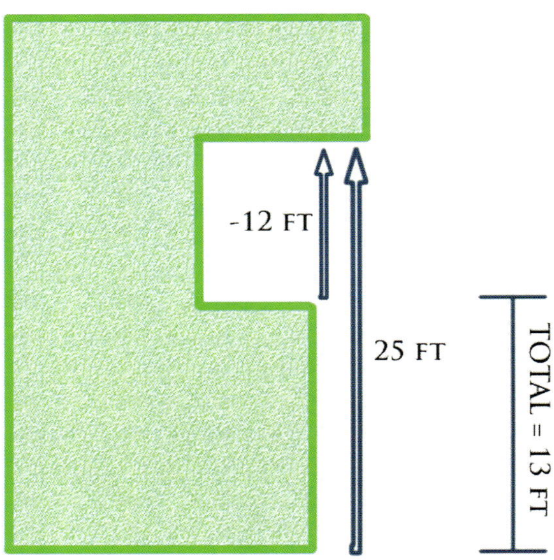

A.06-B Subtractive Method

When measuring a span outside a home where there are no obstructions to shine the laser against (downspouts, columns, etc), measuring a greater span and subtracting the difference is a great time-saving technique.

A.06-C Anchor Method

When measuring inside a home that is more 'beautiful' than normal ('beautiful'=unnecessarily complicated), you may be ready to go insane trying to map out an upper floor. The upper floors are where the architect may have put all the cake-topping 'magnificence' ('magnificence'=unnecessary dormers, cantilevers, bump-ins and bump-outs, and of course the circular turrets and cupolas). And of course, you can't measure the upper floors from the ground outside. Never fear!

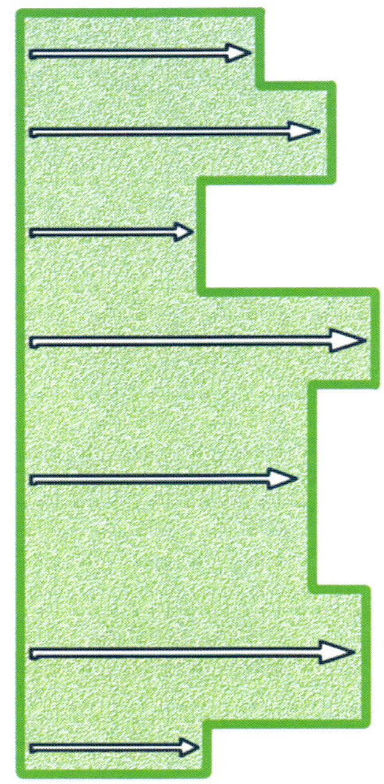

If you can find a wall that runs in a straight line through the floorplan, you can use it as an anchor to measure each crazy change in the perimeter. Since the anchor wall never moves, your potential for error in your measurements is greatly reduced.

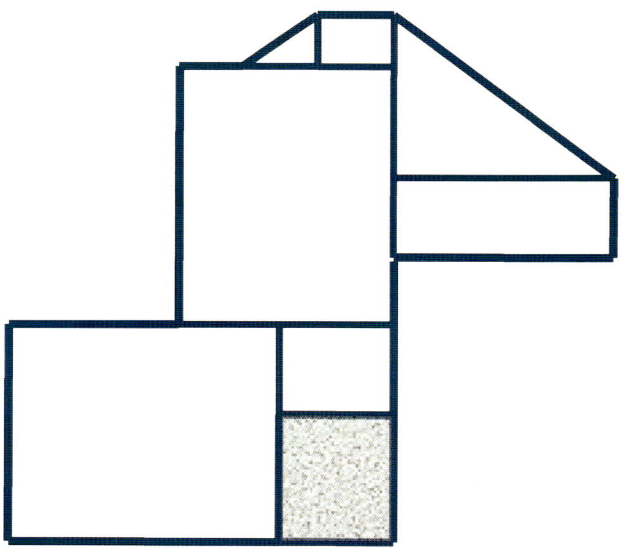

Breaking It Into Pieces

When mapping out a floorplan to measure it precisely, remember that even the most 'beautiful' home is just an arrangement of rectangles and triangles *AS LONG AS YOU DRAW TO SCALE.* Geometry is on the next page.

Converting Inches to Feet

In case you don't have a laser that will measure in hundredths of feet, here's a quick conversion from inches to foot decimals (since you want to do math with the measurements later, always measure in feet, never in inches).

Inch	1"	2"	3"	4"	5"	6"	7"	8"	9"	10"	11"	12"
Feet	.083	.167	.250	.333	.417	.500	.583	.667	.750	.833	.917	1

SQUARE
AREA = S^2

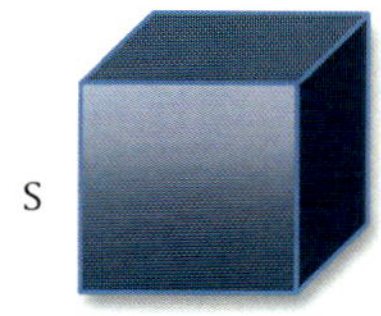

CUBE
VOLUME = S^3
SURFACE AREA = $6S^2$

RECTANGLE
AREA = WL

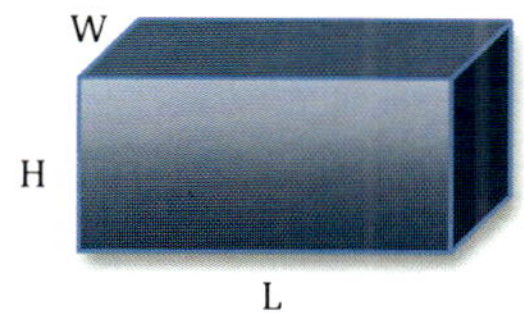

RECTANGULAR PRISM
VOLUME = WHL
SURFACE AREA = 2(HL+WL+HW)

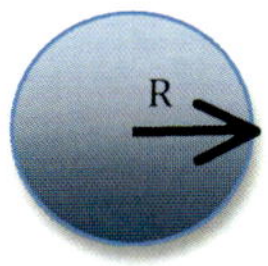

CIRCLE
AREA = πR^2
CIRCUMFERENCE = $2\pi R$

CIRCULAR CYLINDER
VOLUME = $\pi R^2 H$

CURVED SURFACE AREA:
CSA = $2\pi RH$

TOTAL SURFACE AREA:
TSA = $2\pi RH + 2\pi R^2$

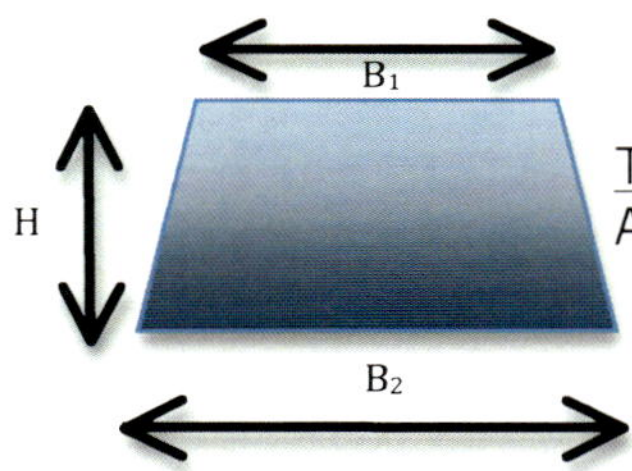

TRAPEZOID
AREA = $\frac{1}{2}(B_1 + B_2) H$

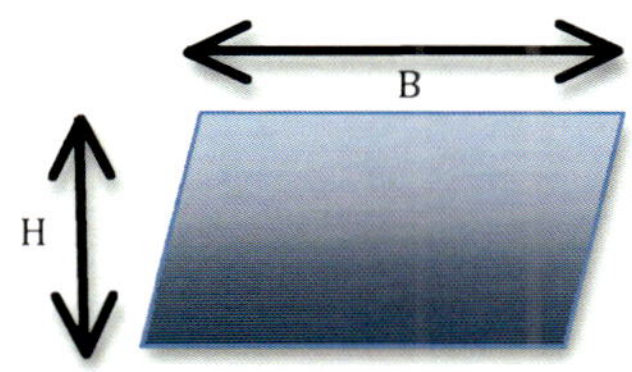

PARALLELOGRAM
AREA = BH

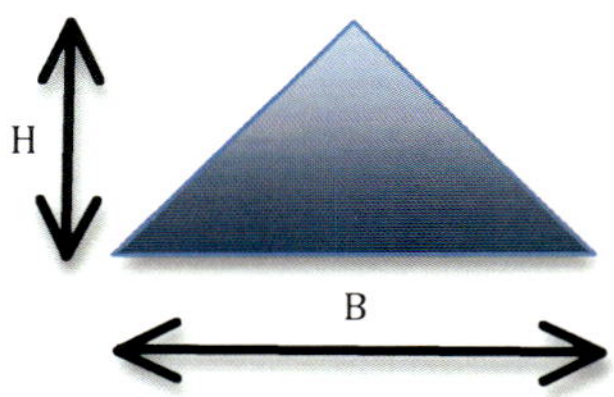

TRIANGLE
AREA = $\frac{1}{2}$ BH

SUM OF INTERIOR ANGLES:
$a + b + c = 180°$

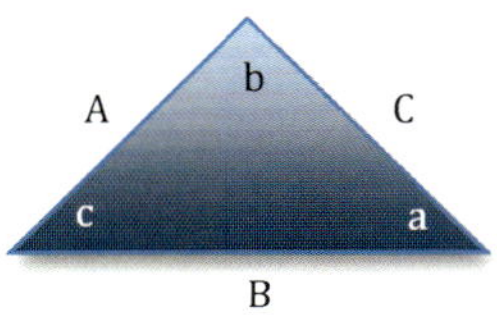

LAW OF SINES:
$$\frac{\sin a}{A} = \frac{\sin b}{B} = \frac{\sin c}{C}$$

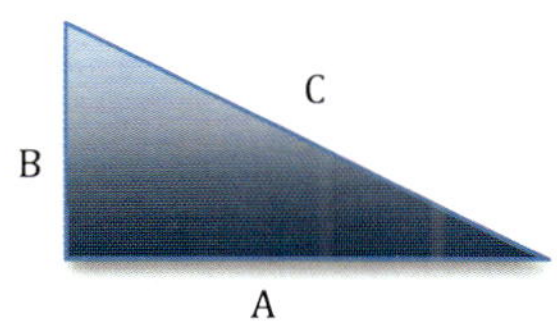

RIGHT TRIANGLE

$A^2 + B^2 = C^2$

A.07 Barometric Pressures Based on Elevation

*Weather also affects barometric pressure; this is for approximation only.

Elevation [ft]	Barometric Pressure [inHg]	Elevation [ft]	Barometric Pressure [inHg]	Elevation [ft]	Barometric Pressure [inHg]	Elevation [ft]	Barometric Pressure [inHg]
0	29.92	**1850**	27.97	**3700**	26.13	**5550**	24.39
50	29.87	**1900**	27.92	**3750**	26.08	**5600**	24.34
100	29.81	**1950**	27.87	**3800**	26.03	**5650**	24.29
150	29.76	**2000**	27.82	**3850**	25.98	**5700**	24.25
200	29.71	**2050**	27.77	**3900**	25.94	**5750**	24.20
250	29.65	**2100**	27.72	**3950**	25.89	**5800**	24.16
300	29.60	**2150**	27.67	**4000**	25.84	**5850**	24.11
350	29.54	**2200**	27.62	**4050**	25.79	**5900**	24.07
400	29.49	**2250**	27.57	**4100**	25.74	**5950**	24.02
450	29.44	**2300**	27.52	**4150**	25.70	**6000**	23.98
500	29.38	**2350**	27.46	**4200**	25.65	**6050**	23.93
550	29.33	**2400**	27.41	**4250**	25.60	**6100**	23.89
600	29.28	**2450**	27.36	**4300**	25.55	**6150**	23.84
650	29.22	**2500**	27.31	**4350**	25.51	**6200**	23.80
700	29.17	**2550**	27.26	**4400**	25.46	**6250**	23.75
750	29.12	**2600**	27.21	**4450**	25.41	**6300**	23.71
800	29.07	**2650**	27.16	**4500**	25.36	**6350**	23.66
850	29.01	**2700**	27.11	**4550**	25.32	**6400**	23.62
900	28.96	**2750**	27.06	**4600**	25.27	**6450**	23.57
950	28.91	**2800**	27.01	**4650**	25.22	**6500**	23.53
1000	28.86	**2850**	26.96	**4700**	25.17	**6550**	23.48
1050	28.80	**2900**	26.91	**4750**	25.13	**6600**	23.44
1100	28.75	**2950**	26.86	**4800**	25.08	**6650**	23.39
1150	28.70	**3000**	26.82	**4850**	25.03	**6700**	23.35
1200	28.65	**3050**	26.77	**4900**	24.99	**6750**	23.31
1250	28.59	**3100**	26.72	**4950**	24.94	**6800**	23.26
1300	28.54	**3150**	26.67	**5000**	24.89	**6850**	23.22
1350	28.49	**3200**	26.62	**5050**	24.85	**6900**	23.17
1400	28.44	**3250**	26.57	**5100**	24.80	**6950**	23.13
1450	28.39	**3300**	26.52	**5150**	24.75	**7000**	23.09
1500	28.33	**3350**	26.47	**5200**	24.71	**7050**	23.04
1550	28.28	**3400**	26.42	**5250**	24.66	**7100**	23.00
1600	28.23	**3450**	26.37	**5300**	24.62	**7150**	22.95
1650	28.18	**3500**	26.32	**5350**	24.57	**7200**	22.91
1700	28.13	**3550**	26.68	**5400**	24.52	**7250**	22.87
1750	28.08	**3600**	26.23	**5450**	24.48	**7300**	22.82
1800	28.02	**3650**	26.18	**5500**	24.43	**7350**	22.78

A.08 Unit Conversions for International Use

Temperature:

Fahrenheit to Celsius: $°C = \frac{(°F - 32) \times 5}{9}$

Fahrenheit to Kelvin: $K = \frac{(°F - 32) \times 5}{9} + 273.15$

Celsius to Fahrenheit: $°F = \frac{°C \times 9}{5} + 32$

Celsius to Kelvin: $K = °C + 273.15$

Length:

Centimeter to Inch: $1\ cm = 0.3937\ in$

Meter to Feet: $1\ m = 3.2808\ ft$

Area:

Centimeter^2 to Inch^2: $1\ cm^2 = 0.155\ in^2$

Meter^2 to Feet^2: $1\ m^2 = 10.763\ ft^2$

Volume:

Liters to Feet^3: $1\ l = 0.03531\ ft^3$

Meter^3 to Feet^3: $1\ m^3 = 35.3147\ ft^3$

Liters to U.S. Gallons: $1\ l = 0.26417\ gal$

Meter^3 to U.S. Gallons: $1\ m^3 = 264.172\ gal$

Density:

Kilograms per Meter^3 to Pounds per Foot^3: $1\ kg/m^3 = 0.0624\ lb/ft^3$

Velocity:

Meters per Second to Feet per Minute: $1\ m/s = 196.850\ fpm$

Meters per minute to Feet per Minute: $1\ m/min = 3.2808\ fpm$

Meters per Hour to Feet per Minute: $1\ m/hr = 0.05468\ fpm$

Flow Rate:

Meters^3 per Second to Feet^3 per Minute: $1\ m^3/s = 2118.88\ CFM$

Meters^3 per minute to Feet^3 per Minute: $1\ m^3/min = 35.3146\ CFM$

Meters^3 per Hour to Feet^3 per Minute: $1\ m^3/hr = 0.58857\ CFM$

Liters per Second to Feet^3 per Minute: $1\ l/s = 2.1188\ CFM$

Pressure:

Inches of Water Column to Pascals: $1\ iwc = 249.088\ Pa$

Pounds Per Square Inch to Pascals: $1\ psi = 6894.757\ Pa$

Millimeters of Mercury to Inches of Mercury: $1\ mmHg = 0.03937\ inHg$

Various:

Air Leakage Metrics:

$$1 \frac{l}{s \times m^2} = 0.19685 \frac{CFM}{ft^2}$$

$$1 \frac{m^3}{hr \times m^2} = 0.054681 \frac{CFM}{ft^2}$$

AUTHORS NOTE:

When using this unit conversion guide, it is recommended that field measurements be recorded in the units customary to the geographical location of the test performed. Once all diagnostic measurements are complete, these measurements should be converted to the appropriate units required for analysis *prior* to applying any equations found in *Home Performance Diagnostics.* It is recommended to carry units throughout all calculations, including the final result, to ensure compatibility and minimize error.